SYSTEMATICS
IN
PREHISTORY

Robert C. Dunnell

Systematics in Prehistory

ISBN: 1-930665-28-8

Library of Congress Control Number: 2001089441

THE BLACKBURN PRESS
P. O. Box 287
Caldwell, New Jersey 07006
U.S.A.
973-228-7077
www.BlackburnPress.com

PREFACE
to the 2002 Printing

...my wife Margaret...suggested that I read a book by Robert Dunnell. However with a title of *Systematics in Prehistory* (Dunnell 1971), it didn't seem to have much to do with evolutionary theory nor did it seem relevant to my interests in paleolithic prehistory and Pleistocene environments. As I have long since learned, however, I should have heeded her advice. (Barton 1997: iii).

Systematics in Prehistory was written over thirty years ago by someone only recently out of graduate school and completely naïve to the ways of academia. As one of my new colleagues at the University of Washington put it: A hardbound archaeology book—what a novel idea. Indeed, books about archaeology, rather than books about the archaeology of particular places, were rather novel. Taylor's *A Study of Archaeology* (1948) was pretty much an isolated analytic effort (one that got its author ostracized from mainline professional circles for decades). Watson, LeBlanc, and Redman's *Explanation in Archeology:An Explicitly Scientific Approach* (1971) was written at the same time as *Systematics* and without knowledge of *Systematics* or vice versa. While it claimed linkage to Taylor's work, this claim is best understood as a legitimizing myth rather than any intellectual connection given the radical differences in approach. *Explanation in Archeology* (EA) and *Systematics* differed in most other respects. EA advocated, in fact largely anticipated, a particular approach to archaeology that the authors dubbed "scientific." *Systematics*, on the other hand, was an effort to explain archaeology as it then existed. And unlike the Taylor book a generation earlier, *Systematics* focused on trying to account for why archaeology actually "worked" in those few areas where it did work demonstrably rather than how it failed to meet new expectations—anthropology in the Taylor case and "science" in the EA case. Furthermore, *Systematics* took as its target the "language of observation," the generation of "kinds," of "facts," a "bottom up" sort of approach to understanding archaeology. The others took a rather different, top-down approach proceeding from assumptions about what answers should look like and thus focused on the "language of explanation." And for the most part, the language of explanation was drawn then, as now, from Western common sense in archaeology. As a result EA went on to become the textbook for the "new archaeology" (later and more distinctively, processual archaeology). Even the Taylor book enjoyed some belated success in that role. *Systematics*, on the other hand, while it did well enough in the bookstores, remained something of an enigma as suggested by Barton's initial reaction related in the quotation, not embraceable by the new archaeology and clearly not conformable with culture history that was the rhetorical target of processualism. It was regarded as difficult to read, hard to understand, and/or of doubtful practical utility. People were not yet ready to question the "objectivity" of science nor the "reality" of "facts." It was in many respects frightening and still is.

Yet the argument that *Systematics* makes is a simple one, germane to all science. Investigators must separate the categories they use to make observations from "natural kinds," or empirical kinds, if such can even be said to "exist." We cannot know the world apart from the templates we use to observe it, a view that even Taylor would come to embrace (1968). The only choice we have is to be aware of the templates and therefore their effects on "facts" or not. When we opt for the latter, the claim is always one of natural kinds, that x is so "because it is real." This is precisely the rationale for folk kinds as well. On the other hand, if the existence of observational templates is admitted, then and only then does the possibility of controlling their effects arise. Only when we know the template used can the artifacts of observation be separated from the phenomenological world. Theory, specifically what *Systematics* calls "formal theory," exists to do this work, explicitly creating a language of observation to generate "facts" that can be explained by the processes and mechanisms of a particular theory. This lies at the heart of the "group"/"class" distinction that is the core of *Systematics:* Groups are sets of things; classes are categories for things. The bulk of *Systematics* is a review of the means available for the construction of categories, the choices made by archaeologists when successful, and even anticipating the quantitative course that archaeology would pursue later in the 1970s and 1980s. The key point is, however, that the purpose, the function of classes is to create groups that can be explained, that are meaningful within some theory. As Richard Lewontin (1974: 8) explained to biologists shortly after *Systematics* was written: "We cannot go out and describe the world in any old way we please and then sit back and demand that an explanatory and predictive theory be built on that description." That he needed to make this point to zoologists is telling. Scientists, real scientists, get trapped by the objectivity myth from time to time, even though their daily practice and disciplinary history belie it.

Failure to recognize this point has led archaeology into a fatal spiral. *Ad hoc* categories were drawn from English common sense. Archaeology was thus compelled to adopt a reconstructionist methodology that conceived the archaeological record in everyday terms. Common sense, our own culture's implicit values and conventions, could then explain it. Reconstitution required creating "attributes" that were not empirical, and archaeology lost its testability. Testing was replaced by examination of how a conclusion was reached. Quantification, as foreseen in *Systematics*, came to be used to create qualities instead of describing variability. Not surprisingly processualism failed.

The gap left by the collapse of processualism as science was filled not by a scientific archaeology but the new relativism sweeping social science (for similar reasons). Yet almost ironically postprocessualist archaeology, as this newest archaeology is commonly known, is likewise tethered to the core thesis of *Systematics*. In attacking the "objectivity" of science, postprocessualists could have drawn heavily on *Systematics* to demonstrate their point. *Systematics* is nothing if not a debunking of the "objectivity" of science, all science, by demonstrating that its "facts" are constructions. The crucial distinction is not between objective and subjective, however, but between explicit and implicit formulation. Consequently postprocessualists miss the methodological point of *Systematics*: Recognizing the constructed nature of kind allows one to *control* the construction, to be able sort observation into artifact and "reality." Rather they glory in being victims of their own unremarked delusions.

Perhaps the impact of *Systematics* would have been more far reaching earlier had it not come out of archaeology. Archaeology is not exactly a place a biologist or a chemist might look for insight into the working of science, especially in the early 1970s. Certainly science has moved in the directions foreseen by *Systematics* since then, even if the social sciences have not, yet everything from the use of cladistics to the "science wars" points out the need to take up the issues addressed in *Systematics* anew.

Looking back, *Systematics*' greatest failure lies in its abstractness. This is not to fault what is there or how it was presented, but to lament that it was not firmly articulated with a particular explanatory theory. Consequently while one discusses how attributes are generated or types formed, there was no content to allow one to actually select attributes or form types. But then there was no explicit explanatory theory in archaeology at mid-century. Language of observation decisions were entirely implicit. Culture history did employ some elements of explanatory theory, but they were hardly explicit. Indeed, I ended up spending a goodly portion of my life trying to extract them from the descriptive literature. New archaeologists openly questioned if theory were necessary, passing that buck conveniently to anthropologists. My own commitment to evolutionary theory lay years in the future, contra to the implications of the opening quotation. Yet in retrospect I could never have come to evolutionary theory without first having understood unit formation. It may be that the resistance to evolution lies as much in failure to internalize the central thesis of *Systematics* as it does with any of the more usual excuses.

Systematics is not an easy "read" (to use the modern jargon) but that is only partly my fault. I have to own up to not having controlled the analytic tradition that did exist for such endeavors, but then my discipline did not encourage such analyses. Anthropological graduate curricula did not include these tools nor guide the archaeological apprentice in that direction—a deficit that I resent to this day and one that I attempted to correct in my own thirty-year teaching career. But many of the distinctions and concepts with which *Systematics* toyed were new, at least in the web in which they were presented; there was, and still is to large degree, no well-oiled path to be shared by analyst and reader when building languages of observation. Finally, challenging the "objective" "existence" of "facts" that can be known through application of "correct" procedure has always been, and remains, a difficult business, crucial as it may be not only to science but to people. It is almost as if people do not want to own up to being sentient creatures. This is what I suspect we mean when we say something is intrinsically hard to understand—something that questions the very tools we use to know. And this is really what *Systematics* is all about.

Barton, C. Michael
    1997   Preface. In *Rediscovering Darwin: Evolutionary Theory and Archeological Explanation*, edited by C. M. Barton and G. A. Clark, pp. iii-v. Archeological Papers of the American Anthropological Association, No. 7. Arlington.

Lewontin, R. C.
    1974   *The Genetic Basis of Evolutionary Change*. Columbia University Press, New York.

Taylor, Walter W.
    1948   *A Study of Archaeology*. Memoir of the American Anthropological Association No 69. Menasha
    1968   The sharing criterion and the concept culture. In *American Historical Anthropology; Essays in Honor of Leslie Spier*, edited by C. L. Riley and W. W. Taylor, pp. 221-230. Southern Illinois University Press, Carbondale.

Watson, Patty J., S. A. LeBlanc and C. L. Redman
    1971   *Explanation in Archeology: an Explicitly Scientific Approach*. Columbia University Press, New York.

CONTENTS

v

LIST OF ILLUSTRATIONS

PREFACE

In evaluating introductory and higher level courses in archaeology, one is struck with the absence of any general text which treats the units employed by the discipline, though all texts are cast in terms of peculiarly archaeological units. Many, if not most, prehistorians have acquired the terminologies by academic osmosis, having been exposed to them over long periods of time first as undergraduates, then as graduate students. But the inconsistencies in the literature of the discipline—the downright isolation displayed in almost all archaeological writing—bespeak the failure of this kind of learning process.

It was in this context that I undertook to write this book, not only for students, but also for myself and my colleagues. It is fair, I think, to call it a first attempt, and I fear in places this is all too clear for comfort. In spite of the predictable shortcomings of such a foolish venture, there is, I think, much to be gained from the attempt itself—not the least of which is to stimulate a more thoroughgoing and deeper consideration of certain basic issues that we, as archaeologists and students of achaeology, all too easily slide under the academic rug in favor of the more active and glamorous aspects of the discipline.

The impression may be given by the pages that follow that there has been no over-all and systematic treatment of classification and unit formation in prehistory. This is not true—but there has been very little. Such treatments, however, focus upon how things ought to be done or what may become practice, rather than what has been done and what has been practiced. This limits the utility of such approaches in teaching as well as in making use of the bulk of the literature. I have tried here to bridge the growing gap between old and new archaeologies by attempting to clarify the old.

In this endeavor, I have been aided by many people, often unwittingly on their part as they simply asked the right question at the right time in the right way. Of the many who have made contributions in this effort, I should particularly single out for special acknowledgment Professors Irving B. Rouse and K. C. Chang of Yale University, who taught me most of what I know of prehistory and who both read my preliminary outlines critically and encouraged me to complete the endeavor. Professor Chang further read the manuscript in draft. His comments are gratefully acknowledged. Professor Michael Owen of the University of Washington read the first half of the book in detail providing the perspective of linguistics. A great deal of credit needs to be given to the students of Archaeology 497 who, over the past three years, have been the demanding proving-ground for much of what is contained within these covers. Without their questions and an insistence upon a straightforward answer, this would have been a far more difficult task than I otherwise could have undertaken. Mr. William E. Woodcock of The Free Press offered encouragement and advice and the kind of willingness to aid that lightens any load. Finally, I should like to thank my wife, who willingly undertook much of the drudgery, editing, and preliminary typing that made this a possible endeavor. Mrs. Carolynn C. Neumann typed the final draft of the manuscript and generously applied her editorial skills. To all these and many more this book owes its existence.

Robert C. Dunnell
University of Washington

INTRODUCTION

Man has probably had an interest in his past as long as he has been man. Depending upon which authorities one reads and which criteria he uses, this interest has been expressed as archaeology in Western Civilization variously—since the birth of that civilization in the Near East, since the time of classical Greece and Rome in the Mediterranean, or since the European Renaissance. Over this period of time—be it five thousand or five hundred years—there naturally have been radical changes in the approach and nature of archaeology.

Today, judging by the meager perspective that can be gained contemporarily, we seem to be entering such a period of change. Often this change is phrased in terms of different approaches or competing schools called the "new archaeology" and the "old archaeology." The "new archaeology" has a different view of the relevance of man's past to his present; its goals appear to be aimed at explanation of man's past, not just at its recitation. With new aims have come, at least to some degree, new means for accomplishing them. The newly envisioned goals provide a clarity of purpose, and the people practicing the "new archaeology" are more systematic and articulate about what they are doing, how they are doing it, and, most importantly, why they are doing it. In looking back, or rather across, to the "old archaeology," the complaints of the new are not so much that the old is wrong—indeed, the old has produced nearly all that we now have of man's past—but that its goals are too narrow, when it has goals at all. An interest in the past is no longer deemed a justification for a discipline in terms of "current relevance."

In particular, the new has criticized the old as being "an art." This criticism has been drawn for nearly twenty years, usually by pointing out that there is no means within archaeology to rationally evaluate its conclusions. One has to be content with "believing" or with assessing the merits of a set of conclusions by a knowledge of the professional status of the individual who did the work.

There is no denying that this was true and continues to be true of much that is done in archaeology and that this is not a healthy state of affairs. Because of these rather obvious faults, there is a strong tendency to reject the "old archaeology" and to replace it, or attempt to, with the "new archaeology." This, however, it to deny the results of the old and, indeed, the "new archaeology" itself which is born of the old and covertly contains much of it.

The practitioners of the old are not without criticism of the new. While their goals may often be appreciated, the "new archaeologists" are brought to task for ignoring priorities of operation, for moving ahead too fast without the proper foundations to bear their conclusions. A good case can be made that much of the laudable effort on the part of the "new archaeology" has been wasted, for it has been based in enthusiasm rather than reason. The tendency to reject *in toto*, or nearly so, the old, has denied the new the experience gained by the old. In the rush to become a science and to produce explanation, the route to science has often been forgotten. Science is not built of novelty. New systems do not appear with each new Ph.D., but, rather, progress is the process of building upon what has already been learned.

Two important products seem to be emerging from the "new archaeology." The first is a very important distinction between field work, the collection and excavation of rocks, and what is done with the rocks after they have been recovered. In short, an academic discipline is growing out of what was once a technical field dealing solely with things. This division has been incipient in archaeology for a long time, but it is the "new archaeology" whch appears to be bringing this into fruition. The distinction between field work and inquiry into man's past will play a major role in the development of what is now called archaeology and will give direction to this development. Indeed, the distinction is a necessary one if explanation is to be achieved. Relegation to a secondary role of that aspect considered by many as the real "meat" of archaeology has undoubtedly contributed measurably to the gap between the old and the new. Because the distinction is important, the technical recovery aspects of the field will herein be called archaeology, and the academic discipline, which is our concern, referred to as prehistory.

The second important and emergent contribution of the new has been its overt search for models with which the discipline may be structured. It is unfortunate that this search has been only partially successful and that the models used have been borrowed from other sources rather uncritically. When science has been employed as the model, the borrowers have looked not to the practice and structure of science, but rather to the philosophy of science, which itself is not a product of science or in many respects an accurate reflection of what science does or how it does it.

Even more detrimental has been the borrowing of models from sociocultural anthropology. This tendency is probably a latent function of the old archaeology, which viewed itself as doing ethnographies of dead peoples and thus as in a dependent relationship to sociocultural anthropology. Looking to sociocultural anthropology for a model to structure prehistory is detrimental because it can only deny prehistory its one virtue, time and change, neither of which is a part of (or can presently be incorporated in a rational way into) sociocultural anthropology. Choosing a model from this source will limit prehistory to untestable "functional studies" executed in terms of differences and similarities rather than change.

The inappropriateness of these and other models to the goals of prehistory has not gone unrecognized by the practitioners of the "old archaeology," and this too has contributed to the division between old and new. While the particular products of this search by the new archaeology can hardly be termed fruitful, the search itself is important and will ultimately shape the discipline in a profitable direction.

The gap between the "new archaeology" and the "old archaeology," insofar as one exists in practice, is in large measure a result of the old, but one which is not being rectified by the new. The problem this book focuses upon chiefly is the failure of the old to produce a comprehensive and overt statement of how and why prehistory works or an explanation of the nature and reliability of its conclusions. There is no general statement of theory in prehistory as an academic discipline. The new, while much more explicit in what it does and how it does it, makes covert use of the old and in doing so suffers from many of the same liabilities.

It is profitable to look at some of the conditions, or causes, that accompany this glaring lack on the part of prehistory, if only to provide some instruction in an attempt to remedy it. It should be clear from the outset that the problem is not a lack of theory, for such is simply inconceivable, but rather the lack of its overt expression in the literature of the discipline. The "cause" most responsible for this omission is the undefined and contradictory usage of the immense terminology employed in prehistory. Like its sister discipline, sociocultural anthropology, prehistory has a tendency to invent a term for its own sake and then argue about what it means for twenty years rather than defining the term in the first place. Some terms are used differently by different authors; other different terms have roughly the same meaning. Much of the confusion and contradiction in prehistory's terminology comes from this source. A given concept has meaning only when it is defined, and once it has been defined, it is an easy matter to evaluate its utility in a given case. The meaning of a term is its definition, not its application, and without a definition a term means nothing and cannot serve as an effective means of communication. Ignoring this has led to the confusing state of the discipline's terms and literature.

Following from this, though certainly meriting some special attention, is a general lack of distinction between the terminology and the referents of the terminology. There is a strong tendency to reify concepts, to regard an idea or a word as the same thing as its referent. Analogously, ideas are not distinguished from observable phenomena. The notion culture, for example, is employed in some literature as if it were a real thing, a huge animal crawling across the planet pulling strings making people do what they do, rather than a concept which enables us to organize the observable phenomena of acts and artifacts.

Closely related to these first two conditions is a general lack of concern with what theory is, again largely as a result of the discipline's preoccupation with the substantive. This does not seem to be a matter so much of theory's being terribly complicated, but rather of its being taken for granted. Methods, for example, are frequently treated as techniques and with little concern for why they work. This has created an enormous problem in prehistory of differentiating between bad methods and the misapplication of good ones. Without understanding why a method works, it is impossible to judge under what circumstances it can be validly employed. This lack of concern with theory has made itself evident in the discipline's terminology. Definitions of concepts quite frequently are formulated for specific problems, but no general concepts are available to consider methods and theory. The many uses of the term "artifact" provide numerous cases in point. Many special definitions are in the literature, yet no general concept of artifact from which these special cases can be derived is in evidence. Thus, not only does the number of terms and meanings for terms approach the number of kinds of problems attended by archaeologists, but means for talking about methods in general, apart from particular problems, are lacking.

The last "cause" which seems to contribute substantially to the malaise of the "old archaeology" is the lack of a clear-cut notion of what prehistory is. More often than not, when prehistory is defined or described it is delineated in terms of its subject matter. Again, the overriding concern for things is evident. When definitions are attempted in terms of goals, these are usually special cases, egocentric definitions of the entire discipline solely in terms of what happens to interest a given individual. They have contributed much to the lack of direction and coherence exhibited by prehistory, something which is frequently obscured from view by the nebulous term "culture history." It is in this respect that the "new archaeology" has not yet made a significant advance, for individual goals are frequently employed to define the field as a whole.

These four "causes" are not really causes, but, rather, are further specifications of the practical problems created by the lack of an explicitly stated theory of prehistory. Historically, these problems relate to the derivation and growth of prehistory as archaeology from a thing-oriented, natural-history stage. Prehistory as an academic discipline, and more particularly as a kind of anthropology, is not very old. Only recently has it appeared in the curricula of universities generally. In the United States, where it is conceived of as a kind of anthropology, it has been forced by its appearance in the academic world further away from things alone, to include ideas as well. In many respects, the "new archaeology" is an attempt to create an academic discipline out of what was largely non-academic endeavor, in the belief that things, so heavily emphasized by the "old archaeology," do not justify the discipline's position in the academic world as a branch of knowledge worth knowing.

The problem which has been outlined is much larger than any one person can seriously attempt to deal with. It is thus not the intent of this examination to cover theory in prehistory exhaustively. Coverage is restricted to the lowest order of theory in any discipline, that of the definition and conception of data, the creation of meaningful units for the purposes of a particular field of inquiry. This is a consideration of the formal aspects of prehistory, the units employed, and the operations performed in arriving at them. It does not attempt to cover the rules by which interpretation and explanation are attempted in the field; indeed, interpretation and explanation lie beyond the scope of the book entirely, save insofar as they have conditioned the construction of units.

The field singled out for treatment here, it must be reiterated, is but a small portion of the kinds of operations and constructions that might be properly called theory. The choice of this particular coverage is predicated on two simple considerations. First, formal operations *must* be performed, covertly or overtly, before any other kinds of operations. One cannot count apples until one knows what apples are, what numbers are, what relations exist between various numbers, and what the point to counting apples is. In the past these formal operations creating the units for the field have been treated, when they have been treated at all, almost entirely covertly, and thus the student has little means to understand this crucial area. And for the professional these simple operations are probably the least well understood of all theoretical matters and consequently are prime contributors to the confusion and misplaced arguments that abound in the archaeological literature. It should be clear from this that the discussion in the ensuing chapters is centered on ideas, "concepts" as they are called, and the operations which create them for prehistory.

The second factor which conditions the choice of subject matter is the current state of affairs within the field. The "new archaeology" is making tremendous, articulate strides in the realm of interpretation and explanation, and it is in these respects that the differences between old and new are most striking. This aspect of prehistoric theory is rapidly becoming accessible to the student through many sources, even in a manner useful at the most elementary level. Most of the units used by the new explanations are, however, still drawn from the old, and often most uncritically. New procedures for unit construction have been proposed, but these have neither made an important contribution nor proved more useful. It is at the level of units that the old and new archaeology are most closely connected, that the old makes its greatest contribution, at least potentially, to the new. The biases in this treatment clearly favor the "new archaeology" in terms of goals and explanation, but are strongly committed to using the units of the "old archaeology" for these purposes. Thus an underlying proposition is that the discomfort created by the formal theory of the old archaeology lies in its implicitness (and thus the possibility of inconsistency and confusion) and its misapplication resulting from lack of problem. The new archaeology promises to eliminate the latter difficulty. This examination hopes to clarify the problem of explicitness.

To the field of concern herein will be applied the term *systematics*, which for the purposes of this discussion is defined as *the set of propositions, concepts, and operations used to create units for any scientific discipline*. A dictionary definition of systematics is not much different except that the word is usually defined in terms of classification and assumes that classification is the way in which the units are created. The definition as phrased here is obviously applicable to all kinds of scientific endeavor; however, our concern will be with those elements which have direct relevance to what has already been done in prehistory.

Within this field of interest the primary goal is to develop a conceptual framework which can be used to understand how and why prehistory works in a formal sense. One thing must be clear: the aim is a conceptual framework, not an operational model. Many different operational models, some of radically different design, are possible within the framework herein developed. This exposition is not concerned directly with how the formal operations of prehistory ought to be done or even how they are done (this is painfully elusive in much archaeological writing), but it is focused on how the formal operations ought to be explicated for evaluation, testing, and comparison. In this respect it is intended to function as a guide to reading what has already been written, by providing a means of correlating and evaluating the divergent literature.

To accomplish this, it is necessary to provide an outline of those criteria that must be met in the formulation of meaningful units for prehistory. In part such criteria are logical operations, but in large measure they depend upon a definition of prehistory. Without this kind of consideration there is no means of identifying nonsense when it is encountered, as it is in all writing from time to time. It has also been necessary to develop a unified set of terms which can be employed as a metalanguage for the discussion of theory in prehistory. It is unfortunate that the subject matter and interests of prehistory are sufficiently complicated or that we understand our subjects so poorly that the metalanguage of mathematics cannot be made to bear the major weight of communication. Whatever the reasons may be, a language which consists of words having only denotations and no connotations must be employed so that we can be certain that ideas are communicated in the form in which they were intended to be understood. The language of theory is *the* crucial item. We can know nothing but words, and in the case of theory it is essential that the words be precise and that this precision can be communicated.

Given the point of view expressed here and the current interests of the discipline, one result, only partially intentional, is the evaluation of some concepts in the role of *expository devices* as more productive than others. Some major gaps in the formal theory of prehistory, presently obscured by vague and conflicting terminology, are exposed as deficiencies. These kinds of evaluations are the natural outcome of systematically examining what prehistory has done and should be regarded as some of the profit that can be gained from this sort of examination. It is important to recognize that such evaluations are bound to the particular point of view and restricted to the particular coverage. The general utility of these evaluations must be established independently.

There is little or nothing new contained in the content of this treatment. It is simply a more rigorous explication of already current notions. All of what is contained herein has found expression many times in the literature of the discipline, though most frequently in a covert manner. However, this is *not* intended to be a literature review. Such is impractical, if not impossible, given the covert expression that systematics has received. Furthermore, a literature review would not be useful since our purpose is not to summarize what has been done, but to analyze it and find out why it works, regardless of why it is said to work. A polling of majority opinion has no place in this kind of approach.

The organization follows logical lines, starting with the most elemental propositions and then deriving those at higher levels. This, of course, is precisely the reverse of the actual derivation of the exposition which began with analysis of the literature and moved from there to the elemental propositions. It might be noted by many readers that symbolic logic, sign theory, and set theory (in specific cases) could have been effectively employed to the ends herein ascribed. The use of general theories of knowledge, however, has been avoided in the exposition where at all possible. This is designed as an introduction to prehistory's theory for students interested in prehistory and not for students of symbolic logic.

To accomplish these general aims the treatment has been divided into two parts. Part I considers systematics in general to provide necessary background for the examination of prehistory in Part II. While Part I treats systematics in this general sense, the considerations are focused on those aspects which are directly relevant to what has been done in prehistory. The initial chapters of Part I set forth the terms and their definitions, and then the later chapters relate the inquiry to the ways in which units can be created. Part II begins by defining the field of prehistory and its relation to the general discussion of systematics. Succeeding chapters consider the ways in which systematics have been employed in prehistory, as well as some of the specific concepts that are the products of these applications. The final chapter in Part II summarizes systematics in prehistory by evaluating the utility of the various kinds of systematics that have been used and the schemes produced, with a stock-taking of where we are and where we can and should go.

In an attempt to provide easy access to the terminology employed, a glossary is appended at the end of the book presenting, by chapter, the terms *introduced* in each associated chapter.

While a bibliography in the ordinary sense of the term is impractical for an exposition of this sort, it is nonetheless advantageous to indicate important sources of directly related materials. Because the subject matter of the first half of the book and that of the second half ordinarily are treated in different bodies of literature, two bibliographies have been provided, one for each part. In these an attempt has been made to include major source materials upon which the exposition has been based, important expressions of divergent views, and examples of the particular subjects treated. Such a listing could quite obviously be extended almost indefinitely, so the brief compilations here are selected works which in the writer's view bear directly on the exposition.

part 1

GENERAL
SYSTEMATICS

1

PRELIMINARY
NOTIONS

*I*t is the intent of the first part of this book to provide a general background in methods available for constructing the formal basis of understanding in scientific disciplines. In essence this involves the construction of a series of linked concepts and assumptions which are usually referred to as theory. Theory, both in prehistory and in the natural sciences, goes much further than what is presented here, for we are not directly concerned with how explanation is achieved, but rather with the formulation of pheomena in such a manner as to be amenable to explanation. Our concern is strictly with formal theory.

This initial background is not correctly assumed to be philosophy of science; instead it is based upon what is done, especially what is done by scientists, and not the way or ways in which non-scientists care to rationalize the procedures. Our model, the natural sciences, is part of Western Civilization and thus largely takes for granted the units by which it operates, much in the same manner (and for precisely the same reasons) that we as English-speakers take for granted the meaning of English words. The sciences, as we are accustomed to use the term, are Western "folk theories" of the phenomenological world, not different in kind or implication from any other pragmatically oriented means of explanation. On the other hand, the formulations of prehistory and other cross-cultural "sciences of man" must be capable of organizing simultaneously both our own system of ideas and things and those systems of the exotic subject matter. Consequently, while the sciences provide a model for the characteristics of units to be used in explanation, the actual construction of units must be considered in more detail than is customary in the sciences. Herein lies the most difficult aspect of making effective use of this sort of study, namely, its familiarity and simplicity. The kinds of things considered are those which all of us do constantly, but *intuitively*. In our day-to-day operation in a single cultural system, the intuitive quality of the way in which we carry out these operations is normally of little or no consequence. There is no need to question, much less any interest in questioning, why a house is a house, because we all conventionally agree on what houses are. The inherent ambiguity is eliminated by our common restricted view of the world; misunderstanding arises infrequently. However, once we turn our concerns to the world as others conceive it, these very operations of deciding what it is that is before us can no longer be taken for granted. The operations must be made explicit for a non-western understanding to be conveyed. In practical terms, this means that to make use of what is presented here it is necessary to rethink or relearn the operations we constantly use to create phenomena, in such a manner as to be able to state how it is we know what we know. The first part of the essay, then, is devoted to providing a general framework for this kind of consideration. It is important to remember that any work of this sort in the social sciences is part of its own subject matter, and to fail to realize this defeats the purpose of the study.

One obvious consequence of this approach to understanding how prehistory works is that we are going to be concerned primarily with words, or concepts as the special words of particular disciplines are called, and the means by which they are constructed. Furthermore, we will also need to concern ourselves with articulating these concepts into a system, a metalanguage in which all meaning is explicit. It is likewise obvious that there is no starting place; one must simply start. Words in ordinary English have to be our touchstone, the means by which the first and most basic concepts are developed. Once beyond the most basic concepts, it will become more and more feasible to create others, based upon the initial steps, without reference to ordinary

English. It is necessary, as an initial step, to create a series of definitions and distinctions which will provide the basic set of terms and meanings to carry the discussion.

Definition versus Description

Pragmatically the most important distinction to be made is that between "definition" and "description." This consideration necessarily precedes all others, for it is necessary to create the basic set of terms. Further, the substantive basis for this distinction between definition and description finds numerous parallels in many of the concepts that follow.

A definition, if one consults a dictionary, is a statement of the meaning or significance of a word. The important feature in the dictionary definition is that definitions have as their subjects words, not objects. Two kinds of definition, differentiated on the basis of how definition is accomplished, are often and usefully recognized: extensional definition and intensional definition. Extensional definition for any given term is accomplished by listing all the objects to which the term is applicable, or doing this within some specified and restricted set of boundaries. For example, an extensional definition of the word "dog" would be comprised of a listing of all dogs, past, present, and future. Clearly, extensional definitions are practical only with a specified set of boundaries, for example listing all living dogs, or all dogs in the state of Georgia, and so on. The only practical application of extensional definition, definition by example, is within some otherwise defined field of time and space. Extensional definition will permit the identification of all dogs as dogs within the restricted realm of living animals. It does not, however, convey what a dog is, those things which go to make up the quality of "dogness." Extensional definitions focus on defining a term in relation to the objects to which the term is applicable. As a result, such definitions are restricted in their utility to defining what is already known. To define the term "dog" extensionally requires that you already know what dogs are in order to make the definitional listing. Ultimately, then, an extensional definition of a term simply means that something is that something because it is, and nothing more. The finiteness of the term's use comes from the necessary restriction of the field to which it is applicable and from which the definition was made. No new animals, for example, can be assigned to the category "dog" if they were not listed as dogs in the first place.

Extensional definitions have considerable utility within single cultural systems in which it is not necessary to know why a dog is a dog because the participants agree on what things should be called dogs and what should not. No information, not already the common possession of the participants, needs to be conveyed. Furthermore, from a pragmatic point of view the worlds of individuals are finite and the number of occurrences of dogs limited to a manageable number, and this provides the temporal and spatial boundaries required for extensional definition. This kind of definition fails, however, in those situations which require conveying information not held in common by the participants or when the referents for the term are not already known and limited in time and space. Such definitions are not suited to the purposes of science or to the kind of consideration made here, because they cannot convey why a thing is that thing, but only that it is.

Intensional definitions, on the other hand, specify a set of features which objects, whether known or unknown, must display in order to be considered referents for a given term. An intensional definition would explicitly list those things which we intuitively use to identify a given animal as a dog, and thus conveys what the term "dog" means in each case of application. This is usually phrased as a statement of the necessary and sufficient conditions for membership in a unit, to which we apply a label in the form of a term or sign. In our dog case, an intensional definition would list a set of attributes which constitute "dogness." Obviously these would not be the sum of all attributes of all dogs, but rather only that combination of attributes which all dogs have in common. If any unknown animal appears, it is readily possible to ascertain whether or not the new animal is a dog simply by observing whether the animal displays those characteristics necessary to be a dog. Thus, intensional definitions have predictive and heuristic value. The particular combination of features which constitute a dog is invariable, and thus provides not only a statement of the meaning of "dog," but also the framework of comparison necessary to establish the relevance of the term to anything which may or may not have been considered when "dog" was defined. It is obvious that intensional definitions are the kind suited to conveying new information rather than simply directing the reader's attention to a portion of what he already knows.

For the purposes of our consideration, definition is to be understood as intensional definition only, and it may be defined as: *the necessary and sufficient conditions for membership in a unit.* This usage will be employed throughout the essay and permits unambiguous understanding as long as a given term is understood only as its definition. For each term developed, a list of distinctive features will be provided, and the term can be used as synonymous with only that particular set of features. A certain danger lies in attributing to a given term characteristics drawn from other usages.

The notion of definition was introduced as one portion of a dichotomous opposition with "description." Description has relevance only for intensional definition, or, rather, it can be readily differentiated from definition only when the definition has been intensional. In our "dog" case, it was noted that some characteristics, those shared by all dogs, are used for the definition. These distinctive or definitive features do not exhaust the attributes of any one dog or any set of dogs. The other attributes of dogs which one cares to distinguish are variable. Some dogs are brown, some are spotted; some bark, some don't; and so on. If one wishes to convey what a given animal is like, once it has been identified as a dog, or if one wishes to talk about one set of dogs after they have been identified, the variable attributes displayed by the individual or individuals under consideration can be listed. Such a listing is what is meant by description. *A description is a compilation of the variable attributes of an individual case or group of cases.* Descriptions can take two forms. They can be simple listings of non-definitive attributes or statements of the frequency of occurrence of non-definitive attributes among the set of cases. Not infrequently the latter kind of description is summarized by listing first the attributes and then the mean and range of the attributes' occurrence, rather than noting each occurrence individually. For example, a description of a set of dogs might note that 14 are black, 17 brown and black, 12 brown, 43 brown and yellow, and five yellow, or this might be rendered in the form of a summary, stating that the coloring of dogs varies from black to yellow and averages light brown. Color for dogs, of course, is non-definitive. To be a dog does not require any particular color, even though experience tells us that those things called dogs exhibit a restricted range of possible colors. Importantly, if a green dog were to appear, it is certain that we would identify it as a dog, but one which was green.

In addition to the variable/non-variable distinction, definition may be contrasted with description in another important manner. While definitions pertain to words, ideas, and other things not phenomenological, descriptions have as their objects only sets of real things. Words, concepts, must be defined; things can only be described. A great deal of confusion can, and indeed does arise from the misapplication of these two devices.

Intensional definitions explicitly identify the invariable attributes required to belong to a unit, so that one can state those attributes which are variable. Intensional definitions provide the framework for description. They establish in tangible terms what is being described and provide the rationale for associating the elements of the description. Intensional definitions are the means of conveying from one person to another the boundaries within which a given description is applicable.

Science

With this background, our first task is the specification of the field of concern for our preliminary consideration, namely science. It is necessary to define this term for inquiry into systematics, and, since prehistory is presented as a kind of science, a rigorous definition sets the parameters for the general structure of prehistory. In English dictionaries the definitions of science, as it is most commonly employed, contain two important elements: (1) it is a kind of study which deals with facts or observations; and (2) it results in a systematic arrangement of facts by means of general laws or principles. The term science is often employed simply for the results of such study, and thus one also finds definitions of science which encompass only systematized knowledge of the physical world. Definitions of science do agree, however, that science is a systematic study involving principles or laws and that it is applied to observable phenomena, resulting in their arrangement as systematic knowledge. On this basis, science can be taken to mean systematic study deriving from a logical system which results in the ordering of phenomena.

Science

One aspect of science not often considered by dictionary definitions is *why* it is done, the purpose to science. Based on observation of what appears to be the case in the natural sciences, the goal of science can be thought of as the explanation of the phenomena considered. What constitutes explanation is something that must be considered to a limited extent, even though explanation *per se* is not of focal concern here. The character of anything is in part determined by the purpose to which it is to be put. Explanation, as ordinarily used, can and does mean many things. Following Eugene Meehan's *Explanation in Social Science*, it is useful to admit two kinds of explanation or goals within science: (1) prediction—a statistical statement of the probability of a given event as the outcome of a known sequence of prior events; and (2) control—a statement of the relationships of a given event to other events and sets of events which enables one to modify the outcome of a sequence to a specified result by altering one or more of the related factors. Using the term "control" does not imply that modification of a given outcome can actually be achieved, but only how such modification could be achieved. For example, a change in the mass of the earth would alter its orbit around the sun in a known manner, though the technical means by which such a change in mass could be effected are not available. Simple prediction, on the other hand, does not tell one why something happens, only that it is probable that it will happen on the basis of past experience. For example, if one smokes cigarettes the chances are very good that one of the several diseases correlated with smoking will overtake the smoker. Yet it is not proper to speak of smoking as a cause of any of the diseases, or any of the diseases as a cause of smoking, because the relationships between them are not known. It is impossible to tell from the correlation alone whether, for example, smoking causes lung cancer, lung cancer causes smoking, or people with a genetic predilection for lung cancer are for the same reason prone to smoke. Nonetheless, the two are linked by a statistical correlation, and thus it is possible to forecast that more smokers will die of lung cancer than will non-smokers. Without a statement of the relationships that obtain between the diseases and smoking, it is not possible to modify the correlation, that is, to alter the forecast that more smokers will die of lung cancer than will non-smokers.

Prediction frequently, but not necessarily, precedes explanation in the sense of control and provides the basis for achieving explanation in this sense. While these two goals of science differ radically, there is one feature of paramount importance which they hold in common. In either sense of the term, the object of explanation is the forecasting or the manipulation of phenomena, and this is achieved by creating classes for the phenomena. In both prediction and explanation some means is required of stating that two things or events are identical in those respects that affect the problem at hand, and this is accomplished by systematics. Systematics functions to convert phenomena into data for a discipline, categorizing historical and time-bound events in such a manner as to create ahistorical units upon which predictions and explanations can be based. This, then, is why sciences are characterized as systematic and as deriving from logical structures. Fully explicated, then, *science is a systematic study deriving from a logical system which results in the ordering of phenomena to which it is applied in such a manner as to make the phenomena ahistorical and capable of explanation.*

Provided with this kind of definition it is possible to differentiate science from other kinds of study, particularly those called history and humanistic studies.

The distinction between science and history is most important, for prehistory is often spoken of as a kind of history, culture history. As a primary goal of the discipline, history is not concerned with explanation in either of the senses employed in sciences. Its purpose by-and-large is a statement of events conceived as unique qualities which happen but once; its primary product is not principles but chronicles, and generalizations based upon them. Because it is not future-oriented and does not attempt to explain events beyond a statement of which events preceded the event in question, it does not have need of systematics. Generalizations demand only associations, statistical correlations of the cancer/smoking sort. History in most of its manifestations has no formal theory beyond the common cultural background of the historian and his reader. Formal theory is not required by history because: (1) it does not have to categorize sets of events into classes since explanation is not an end goal; and (2) the organization of the events is assumed to be known, that is, chronological, and thus the events do not require ordering for the purposes of history. It is for these reasons, especially the lack of systematics, that history is often characterized as "particularizing" (ideographic) and opposed to the sciences which are characterized as "generalizing" (nomothetic). Lest this polemic be misunderstood, it should be remembered that this is a characterization not directly applicable to specific cases. Increasingly, history produces "scientific" results and, conversely, much of anthropology concerned with the

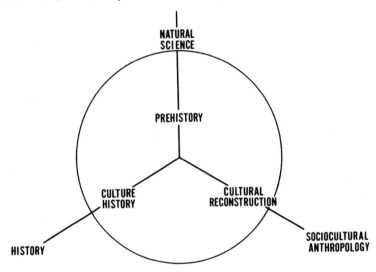

Figure 1. Kinds of archaeology and their relationship to other disciplines.

past, especially the series of results termed "culture history," is strongly historical. There is no neat division in practice (Figure 1).

Complicating a distinction between history and science is the fact that science often makes use of the chronicle. Chronicles in science, however, take a different character because they are oriented toward explanation of what is chronicled, not just its recitation and summarization. In history, the categories used by the chronicle are largely those of the language in which it is written. While chronicles as statements of a sequence of events are common to both history and science, those of science are executed in a terminology which is the product of systematics, whereas those of history are executed in the common meanings of words in the language of the writer.

Humanistic studies are different still and in some cases more difficult to separate in practice from science. In common with history is their lack of concern with explanation of the phenomena considered (though explanation is often given lip-service). In contrast with both history and science, which have overt but different goals, humanities have a diffuse purpose at best. The product, especially in the arts, is often stated to be "appreciation" of the phenomena. Such a product is not knowledge in the ordinary sense of the word, but simply a cultural value. Humanistic studies value things as good or bad in a cultural sense, but not in a pragmatic system with overt criteria for judgment. Thus what is good music varies through time as a style or fad.

Humanistic studies categorize phenomena and so, like the sciences, have their own disciplinary terminologies or jargon. The genesis of these categorizations varies widely from that of science, however. Without a testable product, without explanation, the evaluation of categories becomes a matter of opinion. While the humanities categorize, they do so without the aid of systematics, thus falling victim to accusations of not having any theory. Categorization is done for its own sake rather than for a specified and testable purpose. Even with the great amount of verbiage expended on categorization, humanistic studies still focus on the phenomena and thus retain a strong historical quality as well as a tendency to confuse categories with phenomena (e.g., the popular use of "society" in many quarters today).

The contrast between science and humanistic studies is fairly overt in the case of the arts; however, many of the social sciences are humanistic studies in the sense described here. While they make use of categories and concepts, they do not employ any systematics; categorization is done for its own sake; there is no theory, though the word is often used; the results are not testable or, when they are, the cases are tautological functional studies, and the product is not knowledge but a kind of wisdom which one has to acquire a "feel for" rather than learn. They are contemporary in that what is "good" in social science changes in the same manner as what is "good" in music. Part of the difficulty in separating those non-scientific social sciences from science proper is that a manipulatory purpose is often espoused, and probably has been achieved. The modification of any sequence of events must remain speculative, however, since there is no means by which the unmodified sequence can be posited with certainty. The ability of the non-scientific social sciences to manipulate clearly does not derive from any scientific aspect, but rather from ordinary sources of cultural change. A sociologist is not required to start a revolution; history can tell us that. The intent here is not to deprecate humanistic studies as humanistic studies, only the masquerade of such studies as science to gain crèdibility.

While both humanistic studies and history are readily separable from science analytically, it is often difficult to recognize them in practice. They have been treated here as invariable monolithic entities which in fact they are not. As there are scientific trends in history, so the social sciences include both scientific and non-scientific disciplines, and even within the latter there are scientific practitioners. As in the case of history and culture history, there are kinds of archaeological anthropology that are strongly humanistic, as is most of what could be called "cultural reconstruction" (Figure 1). The main point is however, that the distinctions drawn are not ready-made for practical application, but are polemically designed to limit the realm of practice considered here.

Systematics

As the foregoing discussion of the field of concern here called science should indicate, one thing stands out both as distinctive and crucial: systematics as the means of creating units within a scientific discipline. To avoid this tautology, it is necessary to reconsider in some detail the formulation and the characteristics of units created in science. The consideration will result in a new definition of systematics which is heuristically useful for looking at the ways in which units are created.

All living things respond in a limited number of ways to their environment, and so all things must categorize their environment, sorting it into elements for which they have instinctive or cultural responses. Thus it should be obvious that there are many ways of creating categories and, further, that systematics is best regarded as a special case of such procedures in the larger field of categorization. The object of any sort of unit creation is the categorization of phenomena for one or more purposes, implicit or explicit. There is an infinite variety of ways for men to categorize things, even the same things. Systematics constitutes one such human way of categorization. Units formulated by means of systematics are not held to be "good" or "bad" as are similar kinds of constructions in cultural systems but, rather, are assessed in terms of their use in organizing phenomena for explanation. This latter case can be empirically tested and is usually referred to as "utility." Utility, however, is not an appropriate term, for categories may be "useful" because they are "good" and still have no applicability in organizing phenomena so that prediction and control result. Such empirical testing of units requires a specified purpose for which a set of categories has been constructed. Without such a purpose or problem, testing is impossible because there is no standard against which the organization achieved can be measured for its effectiveness.

Implicit in the discussion is the notion that systematics involves more than a single category; that is, it is a means of creating *sets* of units rather than a single unit. Categorization of any sort involves at least two units: this and everything else. Even in a simple example, it is obvious that the units must be derived from or be analogous to some kind of system; any kind of categorization involves the minimum two units *and* the relationship between them. In the case of systematics the system from which the units are derived is a logical one, that is, one which when articulated involves no contradictory elements and is complete. Furthermore, it is explicit because the relationships that obtain between the categories figure prominently in explanation. So systematics may be thought of as an arrangement of categories, the arrangement being derived from a logical system. Other kinds of arrangement are, of course, possible, and frequently encountered. An arrangement may take the overt form of a system, but upon examination be neither logical nor complete. Many sets of cultural categories are precisely of this sort. Likewise the categorizations of many of the social sciences fall into this pattern. In still other cases no relationships are obvious between sets of categories. This brings into focus another aspect of systematics as opposed to other kinds of arrangements: the system from which the relationship between categories is derived must be explicit. One must not be in the position of having to assume the nature of the relationship between a given set of categories, but rather one must know. If the relationships must be assumed, they will in part be a function of the individuals employing them, thus not replicable and ultimately lacking in ability to organize phenomena in a fashion amenable to prediction or control.

One final aspect of systematics which needs attention is that of the categories themselves. Units produced by means of systematics require explicit definitions. Otherwise, it is impossible to apply the categories to phenomena in a replicable manner. Descriptions have no relevance or, indeed, meaning in terms of categories, since only actual objects can be described. If description were to be employed in systematics, categories could not be employed, and the character of science so based would be historical.

With this further consideration it is possible to define systematics for the purposes herein attended as: *the procedures for the creation of sets of units* derived from a logical system for a specified purpose. We thus can view systematics as a special instance within the broader field of categorization. Categorization is implied in all actions of all living things and therefore its application in arrangement is not restricted to science. Our brief considerations of history and humanistic studies both imply different kinds of categorization. In the first case, pre-existing categories of the language of the historian and his reader, in which all operations and definitions are implicit and which serve to convey meaning by virtue of the shared cultural background of writer and reader, are utilized. In the second, it takes the form of implicitly or, less frequently, explicitly defined categories (often compounded with the phenomena categorized in description) which may or may not be derived from a logical system but share the feature of not being amenable to empirical testing. In actual practice, science may be based upon systematics, but scientists often employ other kinds of categorization. Likewise, social scientists may make use of systematics tangentially, but social science in general is not founded in systematics. In this more than in anything else lies the source of the vast difference in the nature of the results of the sciences and the social sciences.

In the following pages a set of distinctions and concepts is introduced which will provide the basis for considering how systematics operates in science in general. Our goal is not an exhaustive explication but, rather, the explication of systematics in science for the purpose of examining how prehistory works. To this end the preliminary considerations are much simplified from what would be required to thoroughly treat science.

The Fundamental Distinction: Ideational versus Phenomenological

In order to provide unambiguous structure to any consideration, it is necessary for the investigator to be able to separate himself and his tools from the phenomena that he is investigating. The distinction between the phenomenological and ideational is designed to do this. By dividing analytically all "things" into phenomenological and ideational realms a number of important sources of confusion and error can be avoided. It is important to remember, however, that this and any other distinction are artificial. They do not say anything about the real world, whatever that may be; they are designed for a purpose— to facilitate scientific inquiry—and nothing else. Those things which are considered as the *referents for the term phenomenological are those which we can observe,* things and events (e.g., a chair and a solar eclipse). The *ideational* realm is taken to *include those things which have no objective existence,* commonly called ideas. Those things classed as ideational can be known only by means of some phenomenological manifestation (e.g., someone explaining to you by means of noises what he has been thinking). It is not profitable to argue about the relative "reality" of the two categories, for all categories are clearly derived from the ideational realm. We perceive these two kinds of things differently, and thus our means to deal with them are different.

In practical terms, no given instance is purely ideational or purely phenomenological. All phenomena are categorized, and in the process most of their attributes are deleted. All ideas must be given some kind of phenomenological expression before they can be conveyed. Science is designed, however, to enable us to deal with a single one of these categories—the phenomenological realm. For this reason, an analytic distinction between those things which may be observed (things and events) must be clearly separated from those things which cannot (ideas). One important ramification of this distinction lies in the means by which truth may be assessed, for this differs between these cases. In these terms science is a system of ideas used to explain phenomena. By utilizing a distinction between phenomena and ideas, it is possible to separate the means of explanation from the explanation itself. The hard sciences have not much concerned themselves with this distinction at this level. The phenomena they investigate lie at a radically different level than the investigator, and the possibilities of confusion are slight. However, in the case of the social sciences where the investigator is part of the phenomena, the utility of the distinction is much greater. The laws of physics certainly apply to men—but their application is trivial because the level of the laws is far beneath our interest in man.

One encounters this distinction or, rather, a parallel distinction, at lower levels, as the distinction between form and content. Form is analogous to ideational; content, to phenomenological. Forms are not bound to objective existence; they are not real in the usual sense of the word. Form is represented by the categories to which things are assigned. Content is bound to the contingencies of the real world and is analogous to phenomena. Content is represented by the actual things assigned to categories. The content in a given instance is thus unique; the form, recurrent. The barking brown and black dog standing before you at 11:00 A.M., Thursday, April 10, 1957, is unique, for this phenomenon never has been before or will be thereafter; however, dog as a form will recur, as will the barking event, and so on. Implicitly or explicitly, form must precede content, for without it there is no way to identify the content. Unless the forms dog, barking, brown, and so on, were available from English, there would be no way to convey the phenomenon just described, either as a class or as a unique event. The form–content contrast clarifies abundantly the analytic nature of the distinction, for in actual practice the two are inseparable. A former teacher of mine used to employ quite effectively this same set of distinctions at the level of procedures using the terms strategy and tactics: strategy, being a model or plan, is contingency-free and thus analogous to form or the ideational realm; tactics are bound to actual circumstances obtaining in a given case of application and thus are analogous to content or phenomena.

The division of things into ideas and phenomena, into forms and their contents, into strategy and tactics, has an important parallel in the distinction between definition and description. Definitions pertain only to the ideational realm; they are the way in which ideas may be conveyed, even though the ideas and the definitions themselves can be known only as phenomena. Intensional definitions provide a means of circumventing the uniqueness of a given instance by restricting meaning to recurrent attributes and permit the designation of unique aspects as variables. Descriptions, as we noted, are capable of rendering the variable attributes, thus providing content for a form when required, and are bound to a particular set of phenomena, embodying the historical uniqueness of the phenomena described. Descriptions can be made only of phenomena; definitions can be made only for ideas. Descriptions of ideas or definitions of phenomena are nonsensical. We may define our terms, but we must describe our phenomena.

As has already been indicated, the evaluation of things ideational and things phenomenological differs. In the case of ideas, the evaluation is logical, for ideas are neither real nor composed of actual instances. A single idea has no utility, no testability; however, articulated sets of ideas, systems, can be evaluated in terms of their consistency (logical structure), their parsimony (number of assumptions incorporated), and their elegance (simplicity). Only in the case of ideas can one speak of proof. If a system of ideas is logically consistent, that is, there are incorporated no elements which contradict other elements, thus preventing the system from being closed, it is logically true. Of course, this says nothing of its utility, for it may be a trivial truth such as $A + B = C - B = A$, or a nonsense truth with no application. But importantly, no data have any relevance in the evaluation of ideas—the proof of a system of ideas cannot be established by observation, only its relevance to those observations can be so established. Phenomena, on the other hand, may be observed. Being part of the real world, the notions of proof or logical truth are not relevant or useful. The term truth when applied to the phenomenological world, factual truth, is a matter of observation: X event did in fact happen. Future-oriented phenomenological statements are always probabilistic; overtly or covertly, they are statements of statistical probability. In the natural sciences where infinitely large samples of events have been accumulated (e.g., boiling water) the statements made about phenomena are highly probable (e.g., water boils at 212°F at sea level). Because of the high degree of probability, there is a tendency to treat these statements as true in the sense of ideas, which they are not. This predictability is a function of the large number of prior cases, and the distinction between phenomenological and ideational statements must be maintained, especially in the case of social phenomena. Because of the large scale of social phenomena, a large series of prior cases is impossible, and thus the degree of probability that can be attained is lessened proportionately. However, the phenomenological statements of the hard sciences and the social sciences can be of the same kind, only varying in the degree of probability.

This distinction between the ideational and the phenomenological is often phrased in terms of the means of reasoning appropriate to each: demonstrative reasoning in the ideational realm and plausible reasoning in the phenomenological realm. Because ideas are constructed, they have a finite set of specified characteristics which enable them to be completely controlled, completely predictable. Ideas are invariable in the aspects which are of direct concern and maintained so by intensional definitions. Thus logical truth, proof, and demonstration are possible. Phenomena, not being constructed, are infinitely variable and historical and cannot therefore be controlled or anticipated *a priori*. Statements about phenomena must be based on finite sets of prior cases observed, there being no way to incorporate that which has not yet come into being.

The connections between the ideational realm and the phenomenological realm are many. Firstly, we cannot actually deal with phenomena but rather only with categorizations (themselves ideational) of phenomena. Thus two different people seeing the same event see, to a greater or lesser degree, two different events. The common points between the two observations will be in those respects in which they share the same categorizations. Secondly, there is an important connection between the two realms in the form of explanation. Explanation is nothing more than matching a system of ideas whose outcomes and entailments are known (because it is an ideational system) with analogous events in the phenomenological world, thus positing their entailments and outcomes. Modification of the events may be made on the basis of what happens when one or more elements in a system of ideas is modified in a specified manner. Both of these articulations between phenomena and ideas are made by all people as a matter of living and operating in the world. In the case of science, for reasons which have already been discussed, these operations must be explicit, whereas they are more often than not ignored in everyday living.

Insofar as systematics is concerned, the most important articulation between the ideational and phenomenological realms is embodied in the notion of identification. If the goal of science is the manipulation or forecasting of phenomena there must be some means of equating ideational units (classes) with segments of the phenomenological world. Identification is the term applied to this process and is essentially the assigning of real objects or events to the ideational units by means of recognizing attributes of the objects or events that are analogous to the definitive features of the class. While the focus of attention here is upon the construction of ideational units, it must be clearly understood that units so formed are completely useless unless analogous phenomena can be identified with them.

With the foregoing discussion as a basis, it is possible to provide the fundamental notions necessary to construct the examination. A series of concepts will be set forth below to accomplish this purpose. In each case, the concepts can be used only if they are understood as their definitions. If they are considered as having implications or alternative meanings, they will become ambiguous and unable to carry the weight of the examination.

Some Basic Propositions

1. CONCEPT.   Of prime importance is the notion of "concept." This term is used to cover a wide variety of things rang-

ing from a fancy term applied to words which one wants to dignify for one or another reason, to simple ideas or notions. Concept should be understood here to mean the *intensionally defined terms specific to an academic discipline*. The need for concepts is obvious. Academic disciplines have general fields of phenomena in which they are interested with regard to particular kinds of problems. The real world must be categorized in such a manner as to permit the kind of inquiry attempted, not only in terms of specific classes of phenomena but also in relating the level of the classes. One needs, for example, not only the concepts species and genus in biology, but also the terms used to relate these two. The first role of the concept in scientific inquiry is to precisely identify the units being discussed. Secondly, concepts are employed to discuss operations with data and to discuss the theory and method on which the operations are based. Thus in biology one has terms such as evolution, which is a concept of this second category. Because the kind of inquiry is different from what you or I undertake in day-to-day living, the terms must be suited to the task and thus are different from ordinary English. Concepts, then, are words, and nothing more. They are words with explicit intensional definitions which permit the structuring of the world for a specific form of inquiry and which serve additionally to convey the operations performed as a part of the inquiry.

Looking at concepts as words, two kinds may be readily distinguished: those which occur in English and those which are especially invented by a discipline. The first category, those common to a discipline and ordinary English, are the most troublesome. In these cases, the ordinary English word is usually restricted by the academic discipline to *one* of its common meanings. Substantial misunderstanding, particularly by a lay reader, can result by interpreting a particular word in one of the ordinary meanings rather than in the academic sense. The problem is not entirely one of the selection of words. Especially in the social sciences concepts are often borrowed into common English rather than the reverse. Without the strictures of explicit definition that accompany the word as a concept, the meaning of the borrowed word may stray far from the original academic meaning. An excellent example of this phenomenon is the widespread current usage of "society" and "culture."

Those concepts which are words without analogous forms

in common English do not present much of a problem. The layman has to learn a new word and with it the meaning. There is no chance that he will think he already knows what it means since it is not part of his vocabulary. The importance of this understanding of concepts cannot be overestimated in the social sciences. The problem is greatly diminished in the hard sciences where mathematics conveys much of the meaning accomplished with words in the social sciences. Mathematics is a fairly effective barrier against misinterpretation for it consists entirely of symbols which have no meaning whatsoever in ordinary vocabulary. One last thing with regard to concepts ought to be noted: in all cases they are part of the ideational realm. Only the words, as spoken or written, which embody the concepts are ever phenomenological. It is with concepts that science operates, conveying its categories and the operations performed on them, and thus concepts are the cornerstone in understanding the nature of any discipline and its particular inquiry.

2. THEORY. The term "theory," like "concept," is used in a number of widely discrepant ways. The dictionary defines theory in the most common sense as the general principles by means of which a certain class of phenomena may be explained. Importantly, theory is not an explanation, but the principles by which explanation is achieved. Restricting theory to the means of explanation eliminates most of the ambiguity generally involved in the use of theory, for the most common confusion is between the means of explanation and the explanation itself. Explanations are history-bound, necessarily tied to a specific set of circumstances and a finite and stipulated set of data. Theory, on the other hand, to have the power of providing a means of explanation, must be contingency-free, part of our ideational realm.

Theory, then, consists of ideas about general classes of phenomena. The definition indicates that there are essentially two parts to theory: the classes of phenomena and the principles by means of which the classes are related. The principles often go under the label "laws," but to avoid some of the ambiguity associated with "law" we will term the operations and relations between classes principles. It is obvious that both parts of theory are required to produce explanations of anything.

First one must have a set of classes by means of which one can categorize, then identify, and finally convey the meaning of the real world for the purposes to which the theory is directed. This is what is here termed formal theory. Purpose, in general terms, is crucial, for *it is theory that separates the various disciplines from each other, not their subject matter.* Both a physicist and a prehistorian study the same thing—stuff. What is different about the two practitioners is the way in which they care to view stuff, the kind of statements they wish to make about stuff. Both may look at the same piece of stone. The physicist talks about collections and configurations of atoms and can make certain predictions about their behavior. The archaeologist perceives a tool, not a collection of atoms, and the things he can say or is interested in saying about the rock are very different. The two men in their capacities as scientists have seen different things, and only in their common participation in American culture can they share the rock as a rock. The importance of purpose to theory is then obvious—without it there can be no theory, for purpose enters into the conception of the real world. The classes, the categories by means of which the real world is conceptualized, are the first crucial elements of theory. Without these units it is impossible to conceive phenomena as data with any degree of control. As has already been pointed out, the categorization by sciences takes the form of systematics. The units, by virtue of being units, are static entities, and thus the product of systematics is entirely formal.

While ability to categorize the phenomena one is facing is a necessary part of theory, it cannot in and of itself ever generate any explanations, even if accompanied by explicit intensional definitions. Theory must also consist of the relationships that obtain between the units so created. Relationships, not being units, are not formal in the same sense as units. A set of units is not a system until relationships are established between them. If explanation is accomplished by matching a system of known consequences by means of analogs, the relationships constitute a necessary part of theory. Moreover, most theory involves not a single set of units, but many different sets of units, the relationships between which must be stipulated in addition to the relationship between units in the same set.

The relationships, or principles, that articulate units into a system which can be called theory bear a direct relation themselves to the units. It is obvious that the relationships between any two given units must be a function of the characteristics of those units; that is to say, the definitions of the units are the means of deriving the relations between any given set of units. Intensional definitions serve to keep both the meaning of the units and the relationships explicit.

From this consideration the crucial, but partial, role of systematics or formal theory is evident. Systematics must be the beginning point in theory construction, for it is the only means of identifying subject matter. Further, systematics provides the basis for deriving the relations between the units, which in combination with the units permits the generation of explanations. Systematics, on the other hand, is but a relatively small part of what is appropriately regarded as theory, and certainly less visible than relationships or laws. Theory will designate *the system of units* (classes) and *relationships* (laws or principles) *between units that provides the basis for explanation of phenomena.* Our concern here is with the units and their construction.

3. METHOD. A term frequently used in connection with theory is "method," and, like the other terms considered, its usage is varied. This is especially the case since method has many meanings in ordinary English. For the purposes of this consideration method should be taken to mean *a sub-system of a larger theory which is directed toward the solution of a particular kind of problem.* A theory will stipulate or should stipulate all the relationships that obtain between all the units contained within it. When a specific problem is faced by an investigator, not all the theory of his discipline is relevant to its solution. Some segments of it, for his particular problem, will be invariant, and these can be ignored. A method is the model to which the phenomena under consideration will be compared in order to produce the explanation desired. Most frequently, methods can be given the form of a model, and the model can usually be procedural or processual. Not all possible relationships are embodied in the model, but only those relevant to the solution of the class of problems faced. In Figure 2, where the relationship of method to theory is diagrammatically shown, the method seriation does not make use of all the characteristics of all the units used by prehistory, but only those germane to

the problem of chronologically ordering sets of artifacts, and it takes the form of a distributional model.

While theory at least ought to be unitary for a discipline of inquiry, method is not. Even given a specific problem, chronology, there will be many methods for solution, all deriving from the same theory but utilizing different elements. For example, if one of the elements in a particular method for chronology involves stratigraphic position, many cases will occur

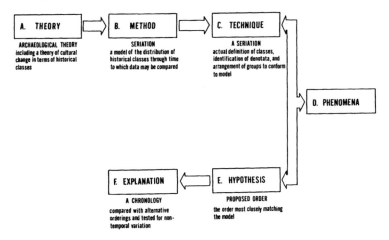

Figure 2. The relationship of the components of scientific inquiry. The problem specified for this example is the derivation of chronology with the method seriation.

in which this variable cannot be stipulated. Other methods are available for chronology which do not make use of this particular variable. A great many models are usually possible for achieving the same goal from the same theory, differing from each other in that they utilize different elements and relationships within the theory to arrive at the unknown which constitutes the problem.

Method, then, is a system directed toward the solution of a particular unit or relationship in the phenomenological world. Its rationale lies in theory and, indeed, method may be considered a sub-system derived from part of a larger theoretical system. Like theory, methods are ideational, not phenomenological. They have no direct relationship to phenomena, but rather provide direction to theory for a specific goal. As we have indicated, methods can usually be rendered as models; however, model and method are not synonymous. Models can be used to convey any system of ideas, both methods and theory as well as other kinds of "abstraction."

While not properly part of a discussion of method as we have defined it here, a consideration of the term "methodology" is warranted by its consistent misuse (in terms of standard English). Methodology is frequently used as a longer word for method, thus being more "scientific." Any dictionary of English is specific in relegating this word to the study of the relationships between various disciplines of inquiry. Methodology is the inquiry into the relationships between the theory of each of the sciences. It is inquiry into inquiry in general, an ideational system designed to investigate ideational systems and not germane to our consideration, nor properly used within any special science.

4. TECHNIQUE. Unlike the other terms thus far considered, technique has seen fairly consistent usage restricted to actual manipulations of data. A technique serves to implement a given method in a specific instance, adapting the method to the contingencies obtaining in the case at hand and satisfying the conditions of application for the method. While methods may be rendered as procedural models, techniques constitute the actual sequence of procedures employed in a case. Actual procedures necessarily differ from the method because they operate upon unique historical cases.

It is through the vehicle of technique that content is introduced into inquiry, and thus technique constitutes the link between the theory and methods of a discipline and the phenomena which they are designed to organize. To fulfill this function, techniques first must order phenomena into meaningful categories. This is the link with theory, for the categories are drawn from theory. Technique applies the definition of those categories to frame the phenomena being considered in terms amenable to the particular method being employed. Identification, from the point of view of formal theory, is one of the most important facets of technique. All methods have conditions under which they are applicable, and the second element in technique is assessing whether or not those conditions are met. It is this step in technique which acts to eliminate mis-application of a method to a body of information for which it is not appropriate. The final step in technique is the actual organization and manipulation of the data according to the stipulations of a given method, with the goal of solving the problem attended by the method. This last step can be clari-fied if you think of an equation in mathematics as a method $(A + B = X)$. A technique in this analogy would be the substi-tution of values for the variables in the equation $(2 + 7 = X)$ and its solution $(2 + 7 = 9)$. The equation itself is just an ordered set of classes and operations for the solution of a par-ticular unknown until actual data are substituted.

The importance of distinguishing technique from method lies in the fact that methods are part of the ideational realm, while techniques, deriving from methods, are part of the phe-nomenological realm, and this means that their respective evaluation is different. Methods are amenable to evaluation in terms of logical truth. They are consistent, simple, and parsi-monious, or they are not. Techniques, on the other hand, once one has evaluated the method involved, are testable in terms of empirical fact. Serious complications can and do arise if method and technique are confused so that methods are evalu-ated as techniques or vice versa. This kind of confusion makes it impossible to distinguish between a faulty method and mis-application of a good method.

Technique is crucial, then, because it is the means of im-plementing theory and methods. Without techniques, theory and method have no utility because they cannot be made oper-ational; they cannot provide explanations of phenomena. Tech-niques permit the matching of a known system in the form of a method with a partially known one, the phenomenon, to pro-duce explanations of the unknown portions of the latter. Tech-nique can be understood as *the application of a particular method to a given set of phenomena.*

5. HYPOTHESIS. The goal of inquiry as we have indi-cated is explanation of phenomena. In science explanation takes the form of hypotheses. *A hypothesis is a proposed ex-planation for a specific set of things or events,* and thus is the product of the application of theory and method by means of

a technique to a given body of data. Hypotheses are probabilistic statements about the relations between phenomena. Hypotheses are not proved; rather, the limits of their utility in terms of prediction and/or control are established. They are replaced by hypotheses of greater utility.

The term hypothesis itself is fairly consistently applied to explanations derived by science, especially those which are held to be tentative. The only confusion lies with the use of theory, as the term "theory" is often applied to statements properly termed hypotheses. Because this confusion between theory and hypothesis is common, and because of the magnitude of confusion that can be so introduced, the relationship between theory, the means of explanation, and hypothesis, the explanation, needs to be treated in some detail. Systematics, formal theory, consists of a *system of units* for the categorization of phenomena into meaningful classes. A method selects sets of relations between some group of units and articulates them into a system within which it is possible to solve for particular unknowns. Techniques, by means of identification, match the units and relations of the method to the partial system of phenomena, and the solution for the relationships or units produced constitutes an hypothesis. Theory is ideational; hypotheses are phenomenological. Theory creates units and the relationships between them; hypotheses recognize analogous units in phenomena and explain the relations between phenomena so conceived.

As with method and technique, theory and hypothesis are not amenable to the same kind of evaluation because they are directed toward different kinds of proof or truth. Theory is amenable to logical verification only. It is evaluated in terms of its elegance, parsimony, and consistency. Hypotheses are amenable to empirical testing only. They are evaluated in terms of sufficiency in addition to elegance and parsimony, under the rubric "scientific method." Regardless of how one derives the hypothesis in a given case (e.g., one may start with a solution and test it or one may "induce" it from the data), the relationship of the hypothesis to the data from which it is derived must be inductively explicated, that is, the data treated as the source of the explanation. Almost inevitably when this is done, not one but several explanations are possible for a given set of phenomena, either as the result of alternative analogs between the phenomena and the classes in a method or as the result of differences in method and technique. Explicating the relationship between hypothesis and data inductively permits the development of alternative explanations, or multiple working hypotheses as they are often called. To approach the explication of the relationship of hypothesis to data deductively, that is, to "test" a hypothesis against a body of data, does not permit this possibility, and, using this means, one finds that one can demonstrate nearly any proposition. One hypothesis may be compatible as an explanation with many bodies of data, but this does not mean that it is the best explanation for those sets of data. The deduction/induction contrast applies here only to the explication of the relationship and says nothing about how the explanation was actually achieved. To muse over the actual derivation of explanations is to predicate science on psychology, something that is neither necessary nor profitable.

Once there is a series of alternative explanations or hypotheses for the relations obtaining between a given set of phenomena, then the familiar form of evaluation, scientific method, is clearly in evidence. Competing explanations are weighed in terms of: (1) their respective elegance, the simpler the explanation, the better the explanation; (2) their parsimony, whether the explanation posits any data not in evidence; and (3) their respective sufficiency, whether the hypotheses explain all of the data. Weighing in these terms will usually eliminate many if not all but one hypothesis, but not infrequently there will still be competing hypotheses. These can be further evaluated by (1) deductively applying them to data from which they are not derived and seeing which explanation has the greatest power of explanation; and (2) by deducing consequences of the explanation and then testing to see if the consequences are in evidence in the data. Even if there are no alternative hypotheses beyond the initial evaluation, the credibility and probability of the hypothesis are enhanced by applying it to data from which it is not derived and by examining its logical consequences in the data available. To complete either type of test, the relation of the data to the hypothesis must be restated inductively so that one can demonstrate not only the sufficiency of the hypothesis but also its elegance and parsimony. The simplest, most parsimonious explanation which encompasses the most cases and which has logical consequences that are verified is best. The temptation, of course, is to regard such an hypothesis as true rather than highly probable or credible. Since, however, its "truth" is predicated on testing against data, it cannot be considered true unless tested against all cases, which is, of course, impossible (future events, etc.).

The ultimate evaluation of a hypothesis lies, then, in its power of explanation of phenomena. It must be tested against facts, and it is the product of this testing that permits the evaluation of the hypothesis. Theory, as a system of ideas, it not testable in terms of facts, for the facts are generated by the theory in the categorization process. This is the genesis of the "don't confuse your facts with your theories" statements. Given that the differences between theory and hypothesis are largely a product of the former being ideational and the latter being phenomenological, they must be evaluated by appropriate means. A means appropriate for one is not appropriate for the other. Empirical testing is not relevant for theory. Logical consistency, on the other hand, is not a "test" to be applied to hypotheses.

The effects of confusion between the two realms of notions can be clearly seen in contrasting principles, laws or elements of theory, and generalizations, statistical abstractions, or aggregates of events. The source of confusion between the two lies in the fact that neither are real, or, as it is more commonly phrased, both are abstractions. Principles, as segments of theory, are ideas; they are not testable in terms of phenomena. They may or may not be relevant in any particular instance (e.g., law governing the relationship between voltage and amperage in electrical circuits and the flight of a bird), but the fact that they do not permit explanation of a given case is not valid evaluation. Generalizations, on the other hand, are statistical models built up from observations. Their statistical quality may be overt or covert, but they are always normative statements based upon a given finite set of cases. Generalizations are in a very real sense nothing more than a set of averaged facts. Generalizations thus change with each new increment of information, and they are either accurate means and ranges of a set of events or they are not. Generalizations are a form of description, a form useful in many kinds of cases (the boiling of water) when their nature as generalizations is appreciated. Far less infrequently than one might hope, generalizations are employed as principles. This effectively nullifies a distinction between ideas and phenomena and eliminates the possibility of rational evaluation. When a generalization, representing a statistical description of a set of past events, is employed to understand new information, it is called prejudice in our social world (I was bitten by a dog once; therefore, all dogs bite). Surprisingly enough, this same procedure is not uncommon in some social sciences today, but without the perspective provided by the social situation. Failure to realize that generalizations are neither explanations nor means of explanation has robbed much of the social sciences of an ability to explain. The results of such misapplied generalization are incapable of evaluation in the manner described above, and these products become matters of untestable opinion. The literary polemic which abounds in the social sciences is possible only because of the lack of definitive means of evaluating statements based upon generalizations employed as theory.

The pragmatic point to making a distinction between definition and description, between ideational and phenomenological realms, is that such a distinction permits the rational evaluation of statements by matching the kind of evaluation to the nature of the statement. When explanation is the goal, as it is in science, rational evaluation is an absolute necessity to establish explanations and to modify the means by which explanation is reached. Figure 2 presents a simplified model of the relationships between the terms used in categorizing scientific inquiry. From the nature of the diagram, it is obviously not intended to convey the actual procedures used in reaching explanation, but rather to be a formal model of how the procedures are logically related to one another. While the example provided by the text in this figure is concerned with a particular problem within archaeology, the general structure is applicable to any kind of inquiry: one starts with a set of explicitly defined notions (theory) which are capable of being organized according to some of the defined relations in a model for the solution of a particular class of problems (method), which in turn is capable of being matched with phenomena (technique) in order to produce a testable hypothesis capable of being used as an explanation (prediction/control). The model, of course, assumes that no modification of the theory and method is required for the solution of the class of problems treated in the figure and does not indicate alternative methods for the solution of the same relationship among phenomena. Were these procedural options included, the model would have to contain more elements and relations, but this is not necessary to illustrate the basic relationships within inquiry.

As indicated previously, the totality of inquiry is not the subject of our examination, but rather the restricted portion we have termed systematics, the creation of units in theory and method. The crucial role played by the products of systematics in the over-all structure of scientific inquiry is evident, for it is with these units that phenomena are apprehensible, that they can be structured by technique to produce explanations. Systematics is the first step in achieving explanation and lies within the ideational realm, though it must be applied to phenomena. This discussion is not intended to apply to how explanations are actually achieved. Anything new is first learned by guessing. The structure to inquiry outlined above is not a program for how to guess, but how to demonstrate the utility of the guess and precisely convey to others the content of the guess.

2

CLASSIFICATION

*T*he word classification is intimately associated with systematics; indeed, the two are often considered almost synonymous. It was noted in the first chapter that the common English usage of systematics implies that systematics is a product of classification. However, since classification often covers a wide range of different devices, it is necessary to define classification. To do this it is useful to consider classification as a special kind of a larger, more inclusive phenomenon which, for lack of a better term, can be called arrangement. It will be possible then to view classification as the kind of arrangement which leads to systematics in science.

Arrangement can be taken to encompass any activity which has as its product an order or orders, any procedure which leads to unitizing. One can talk about arrangement of ideas, thus speaking about arrangement within the ideational realm, and one can talk about the arrangement of things, arrangement applicable to the phenomenological realm. The way ordering is brought about, the nature of the units created, and the order which units display, can serve as the basis of distinguishing kinds of arrangement (Figure 3). It is obvious that arrangement is required for any kind of inquiry or, indeed, any kind of reaction to either ideas or things. As men we arrange things and ideas continuously in daily living, and we do so both overtly and covertly. For the purposes of scientific inquiry, the arranging must necessarily be done overtly so that the arrangement and the rationale for the arrangement can be conveyed. From the utility of distinguishing the phenomenological from the ideational, especially in terms of evaluation, it follows that arrangement may be approached along these same lines.

Classification will be restricted to arrangement in the ideational realm and defined as *the creation of units of meaning by stipulating redundancies* (classes). *Grouping* will be used to denote arrangement in the phenomenological realm and defined as *the creation of units of things* (groups). Grouping and classification are articulated with one another by means of *identifica-*

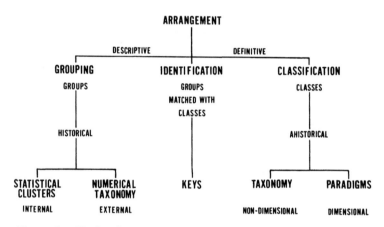

Figure 3. Kinds of arrangement.

tion, the process of using classes to assign phenomena to groups, essentially matching a system of classes with a body of phenomena to create groups which are analogous to classes. Following from the considerations undertaken in the first chapter, it is obvious that classes are *useless,* without groups, and that groups are *meaningless* without classes. In the course of day-to-day living, a distinction between classes and groups is not necessary, for no new information is being conveyed within a single cultural system and evaluation is not overtly conducted; however, for the purposes of scientific inquiry and the evaluation of its results, it is necessary to make such a distinction. Without it evaluation is impossible. The lack of such a distinction in much of the archaeological literature has created a great deal of the confusion in evidence and represents the transfer of a common-sense approach to scientific inquiry.

The products of classification and grouping, classes and groups respectively, contrast as members of the ideational and phenomenological realms. Classes must be defined, cannot be described, and enjoy no objective existence. They are parts of ideational systems. Groups, on the other hand, are not and cannot be defined, but are described and are bound to a given position in time and space. The category "dog" is timeless—an individual case, Rover, is an historical phenomenon. Rover may be described, "dog" may be defined, and if we cite those things which cause Rover to be categorized as a dog we have identified Rover. The important thing is that "dog" and a given dog are two different, but related, things. Since our concern is expressly theoretical, the main concern will be with that aspect of arrangement called classification, but it is well to remember that to realize any utility from classification it must be articulated by means of identification with groups of phenomena.

Classification is directed toward the production of classes, units of meaning. Classes have a special relationship to definition; indeed, in many respects classes are identical with intensional definitions. A class, as a unit of meaning, can be thought of as a conceptual box created by its boundaries. The boundaries are established by stating the criteria which are required, the necessary and sufficient conditions, to be included within the box or class. The set of criteria which determine the boundaries constitutes an intensional definition of the class called its *significatum*. Classes consist entirely of their *significata*. Thus, since a class is created by the criteria for membership, all of its characteristics as a class are known and invariant. There is nothing to be described. The meaning of a class is its *significatum*. If an object, Rover, is assigned to a class, "dog," by identifying in Rover those criteria necessary and sufficient to be a member of the class "dog," then he may be considered a *denotatum* of the class "dog." When another animal is assigned to the class, it is because the new animal displays those same characteristics which caused Rover to be called a "dog." Calling both animals "dog" means that they are identical with regard to the conditions required of "dogs" and thus may be treated the same in any theory employing the class "dog" (e.g., to predict whether or not the new animal will bark or meow when it makes a noise). Referring to Rover as a "dog" means only that he displays the definitive characteristics of dogs and nothing else. Obviously, assigning Rover to class "dog" will not tell one what color he is, though it might restrict the range of possibilities. Color is not part of the meaning of "dog."

It is necessary to distinguish the process classification from classification in the partitive sense, *a classification*. A classification is a system of classes produced by means of classification. The application of classification results in the production of a set of classes, not a single class, which are linked with each other through their *significata*. It is the relationship of the elements in the *significata* of a set of classes that gives a classification its system nature. These links between *significata* are a direct function of the manner in which classification has been done, and they control the form of the classification. Kinds of classification and the linkages that they produce between classes will be treated in the following chapter. Here the focus is on the classes themselves and the general characteristics which are common to all forms of classification.

Classification is best treated, once defined as the process for the creation of units of meaning by means of redundancy, in terms of a series of axioms or consequences of the definition. In doing this, both the assumptions upon which it is founded and the rules for its operation may be made explicit. Choosing some consequences from an infinite field also provides an opportunity to look at classification along parameters which are a source of difficulty in prehistory, and thus to raise some assumptions associated with classification to the level of issues. The five axiomatic, issue-oriented statements below provide the parameters germane to the present task. They are presented in order from the most general to the most specific since the demonstration of one closely follows on the demonstration of the others. Each of these consequences will be considered in detail:

1. *Classification is arbitrary* (a particular classification is not inherent in any field or phenomena).

2. *Classification is a matter of qualification* (quality has primacy over quantity).

3. *Classification states only relations within and between units in the same system* (classification is an organizing device, not an explanatory device).

4. *Classificatory units, classes, have primacy over the labels applied to such units.*

5. *Classifications, classification, and classificatory units have primacy over structures, structuring, models, and model-building.*

1. CLASSIFICATION IS ARBITRARY. From the outset, classification assumes a particular view of the world, in part as a consequence of being a kind of arrangement. The external world is conceived of as an unsegmented continuum of form through the dimensions of time and space. What appears to be unitary and discrete at one level of inquiry is composite at others and component at still others. Classification assumes that the external world, the phenomenological realm, can be most profitably conceived of as comprised of an infinite number of uniquenesses or instances. These instances take on the qualities of things or the qualities of events depending upon the point of view assumed by the viewer. Customarily, things are instances in which the dimension of time is perceptible. The customary division into things and events, however, is just that, customary, and not of general utility. Given this view of the world, it follows that chaos is inconceivable and therefore not a profitable notion in inquiry. Conversely, it is assumed that the infinite number of uniquenesses that constitute the phenomenological realm must have inherent order. If it is assumed that there is *an order*, inquiry takes the form of discovery; however, such an assumption greatly inhibits evaluation, for if *an order* is assumed, it is not possible to discover that multiple orders, or indeed no order, obtains in the phenomenological realm. Following from this, classification rests upon an assumption that the uniqueness of the phenomenological realm is capable of order, but not that any particular order is inherent.

These notions are not to be taken as any kind of ultimate truth, or even true in a more limited sense. They are heuristic devices and nothing more. To begin any reasoned pattern it is necessary to start someplace; the beginning is always assumptive. What is necessary is that the assumptions be made as innocuous as possible so that it becomes irrelevant whether or not they are true. The assumption made by classification, that the external world is composed of an infinite number of uniquenesses which are capable of order, is just an assumption, for it allows for the possibility that there may be a single universal order, several orders, or, in an oblique way, no order at all. Should the latter be the case, chaos is the result and is beyond our apprehension, thus identifiable.

Heuristic devices taking the form of primitive assumptions are necessary for the purposes of science, for only in this manner of conception is it possible to make the foundations of our knowledge susceptible to evaluation. To do otherwise is to assume that which science proposes to demonstrate. In the hard sciences, these basic propositions are of less importance, indeed trivial, for there is traditional consensus about these matters, now largely implicit. They become crucial in cases of constructing new sciences where such consensus has not been traditionally employed or achieved.

Given this conception of the phenomenological world, classification, as the ideational aspect of arrangement, can be viewed in its simplest form as the primary adaptive mechanism of all animate life. It provides the internal means for reducing reality's uniqueness to a manageable number of classes for which a finite organism has responses. It reduces the non-repetitive world to redundancy by stipulating identities and thereby creating classes of phenomena—indeed, creating the phenomena themselves. Looked at temporally, classification introduces the possibility of repetition of events as well as static categories of things.

It follows that classifications can be produced at an infinite number of levels, proceeding from the pole of total uniqueness to the pole of total unity or singularity. Total uniqueness is, of course, chaos which is undefinable and thus not classification. Total unity lumps everything into a single undefinable unit and again is not classification. Systems of units lying between these two poles are all potentially capable of definition and may be profitably considered classification. It is in this proposition that the first element of arbitrariness is introduced into all classification. In order to create a classification, the first step must be the stipulation of scale, the selection of one of an infinite series of scales, at which classes are to be formulated. A more detailed consideration of the notion of scale and related concepts is undertaken in Chapter VI; however, an example familiar to prehistorians is that of deciding what part of a discovery one will treat as artifacts—the site itself, its houses, or their post molds. Of course, all will be treated, but in different classifications. In prehistory, as will be pointed out later, the beginning point is the discrete object and other scales are reckoned from this point. In any case, classifications are always at some specifiable scale of phenomena. Insofar as scale is not inherent but a matter of selection, all classification is arbitrary.

A concurrent step is the subjective selection of the field for which the classification is to be constructed. Classifications never attend the totality of phenomena at a given scale, for, at least from the point of view of the systematist, one would be faced with a field of infinite size. At a given scale, say discrete objects, the field will be narrowed to some "kind" of discrete object. This means that if one is going to create a classification for animals (a kind of discrete object), animals must be defined external to the classification. You have to know what animals are before you can conceive of kinds of animals. Again, because selection is involved and because the definition of the field must lie outside the classification itself, arbitrariness is introduced into classification.

In day-to-day living both field and scale are covert and culturally controlled. The definition of fields and scales for classification are usually accomplished by the theory of a particular discipline and should be explicit to allow evaluation and revision. Further along we shall see that the definition of the concept "artifact" is crucial to prehistory for just this reason—it defines the field (at the level of discrete objects) for classification in prehistory.

Having defined the scale and field which a given classification is to attend, a third arbitrary element needs to be introduced before a classification can be accomplished. This is the discrimination of attributes of the field at a stated scale beneath that of the field. If, for example, one wishes to create a classification for animals, in addition to specifying the scale at which animal is conceived and defining what animals are, one will have to stipulate attributes of animals, parts of animals, which can be used to divide animals into kinds. *An attribute is the smallest qualitatively distinct unit involved in classification.* Essentially two things are involved in the discrimination of attributes: the stipulation of the scale at which they are formulated, and the division of the scale into the intuitive units called attributes.

The definition of the scale is just as arbitrary in the case of attributes as it is in the case of the field. Further, the division into discrete attributes must always be intuitive, for the definitions of the attributes will lie outside the classification. The discrimination of attributes, like the definition of the field, is customarily embedded in the theory of a particular discipline.

The attributes discriminated become the potential criteria for classification, but potential only, because further selection is required. The selection of attributes *as criteria* introduces the fourth and final arbitrary element. Following from the general assumptions made by classification, the attributes possible are infinite, and only a finite and usually very limited number of attributes can be employed in classification. Obviously, not all attributes can be used. Even if it were possible to use "all" attributes, there would be no point to doing so, for the product of using "all" attributes would be the division of the field into an infinite set of unique cases. The net result would be a statement that everything is different from everything else, a non-productive conclusion because this is assumed from the beginning and is certainly not a kind of classification. Which specific attributes are selected is usually controlled by the particular problem for which the classification is designed. For example, if one were interested in animal ecology, one might choose the food-getting habits of animals as the basis for a classification resulting in classes such as herbivores, carnivores, parasites, etc. Another problem dealing with animals, say their evolution, will make use of different kinds of attributes and result in a different set of classes.

In addition to selecting the kind of attribute, the systematist also selects their number, thereby establishing the level of classification. The larger the number of dimensions of attributes used, the more numerous the classes and the finer the discriminations will be. Ordinarily this decision is made with reference to the specific problem being treated; when it is done categorically the lumper-splitter controversy arises.

The discrimination of attributes and the selection of some of those attributes as criteria are frequently lumped as the *analytic step* in classification, because it is in making these procedures explicit that the classification of science differs most radically from everyday behavior. Literally, analysis means to break things down into their component pieces. While this is obviously not what is done in the discrimination and selection of attributes, structurally it is the same procedure, for it is the conception of component pieces. The analytic step is analogous to the "etic" part of the "etic-emic" dichotomy which has gained some currency in anthropology. Analysis accomplished at a scale beneath that of the field lies outside the bounds of classification and forms the basis for classification.

What constitutes analysis and what constitutes classification can be defined only in the context of *a* classification. Analysis (etic) and classification (emic) are relative in a general context. What is analysis at one scale is classification at the next lower scale. For example, one could create a classification for animals based upon locomotive apparatus, and then one could create a classification of locomotive apparatus which would be based on attributes of such apparatus. In the first case the locomotive devices are attributes; in the second they are the classes. This relative aspect of analysis and classification follows directly from the assumption that the phenomenological world is comprised of an infinite series of scales from the pole of total uniqueness to the pole of singularity.

Before leaving for the time being the notions of analysis and arbitrariness it might be well to note one important distinction not clearly evident in the foregoing discussion, namely, that the term "attribtute" is ordinarily used to mean two decidedly different things. First, it is used to designate particular qualities of particular instances. In this sense attributes are unique, non-recurrent, and wholly within the phenomenological realm. Rover in our earlier example is unique in all of his attributes. Rover's color is Rover's alone. Attribute is used also to designate classes of attributes. The color category "brown" applied to Rover's color is an attribute in this second sense. Such classes of attributes as part of analysis are not the product of classification, but are intuitive, at least in relation to the scale at which classes are being formed. In the case of Rover, he has an attribute of color, which is assigned to a class of attributes "brown," which in turn is intuitive insofar as the class "dog" to which Rover himself belongs is concerned. Again, attribute in the sense of some quality of an object or event is different from, though closely related to, the name and category to which that quality is assigned. Hereafter, when discussing these kinds of things in a general context, attribute will be restricted to the unique quality of a specific instance in the phenomenological realm, and feature will be used to designate the classes of such attributes. Because these considerations are pragmatically trivial at this point in the development of the hard sciences, the distinction is not commonly made; however, in the case of prehistory the distinction is crucial and, indeed, has found recognition in two terms for many years.

To summarize this first axiom of classification it should be obvious that the term arbitrary is applied not to unreasoned or uncontrolled decisions and discriminations, but to the specific assumptions that are necessary to begin classification, given a conception of the phenomenological world as posited by classification. *Arbitrariness means only that the discriminations made are not inherent in the phenomenological world as the only distinctions possible.* Arbitrariness is necessarily introduced in all classifications at four points:

A. *The stipulation of the field* to be considered by the classification.

B. *The stipulation of the scale* within the given field at which classes are to be formed.

C. *The stipulation of the attributes of the field,* involving first the definition of scale beneath that at which classes are to be formed and then its division into attributes.

D. *The selection of attributes as criteria,* both number and kind, for defining classes.

The first two elements "locate" where the classes are to be formed; the last two, usually grouped together as analysis or the analytic step, "locate" the means of creating the classes. In the sense employed here, all classification must be arbitrary. No classification can be natural. Arbitrariness inheres in these four sets of decisions which must be made and defined outside the classification itself.

2. CLASSIFICATION IS A MATTER OF QUALIFICATION. Whereas the first axiom attends the assumptive foundations of classification, this second statement focuses on characteristics of the process of classification. The distinction between quality and quantity, between units and counts, follows directly from the initial distinction between ideational and phenomenological and between definition and description. Thus the assertion that qualification logically precedes quantification is simply a more

closely specified case of the priority of definition over description, here in the context of classification. The centuries-old debate among philosophers about quantity and quality is not relevant here, for the terms are defined much differently in that debate. As has already been indicated, a class is created by an intensional definition, by the statement of a necessary and sufficient condition or set of conditions. This axiom thus asserts that the necessary and sufficient conditions are the product of qualification. For the creation of classes, it is necessary that one have a scale and a field for which the classes are to be formed and features beneath that scale and within that field. The features, primitive classes themselves, provide the conditions or criteria for the formulation of classes, and, as has already been shown, this discrimination of features is a matter of distinguishing qualities, not quantities. The manner in which features are employed to create classes varies from one kind of classification to another; however, the definition of a class is always a list of those features which a given thing within the field and at the given level specified for the classification must display in order to belong to a given class. In those cases in which more than a single feature is required, the linkage between features, the means of combining them, is the physical entailment of their analogous attributes in the same object or event. Co-occurrence, then, is the means of linking several criteria for a definition. For example, if a class is defined as "yellow-rough" than all objects assigned to this class as *denotata* must be both yellow and rough, and only those objects both yellow and rough can belong. Objects by virtue of being yellow *or* rough cannot be considered *denotata* of this class.

Given that there is some current interest in arrangement, especially in the biological sciences, and that sets of terms have been employed to talk about arrangements, it might be well to point out that classification is here taken to be "monothetic classification." Within this framework its opposite, "polythetic classification," is not considered classification at all, for it displays some very different characteristics beyond the implied differences in the number of defining characteristics.

The assertion that quality is logically prior to quantity can be examined outside of the context of classification. You cannot count something until you have something to count. More often than not, when emphasis is placed on quantity this means only that the classification which produced the things being counted is covert. Only units may be counted, and units are the product (as well as the input) of classification. It is further obvious that if the units have not been defined prior to being counted, there is no way to know what the count means. If one sets out to count "apples" without any means of identifying apples, one might well end up counting apples, some oranges, and a few red rubber balls. Whether or not rubber balls were counted is largely irrelevant. What is important is that there is no way to tell whether or not there are any red rubber balls included in the count. For some purposes it might not matter, e.g., if you want to know something about roundish red objects; for other purposes it may be decidedly figure, e.g., to be able to distinguish food. Again, there is no way to judge to what purpose the counts may be usefully put unless the units which have been counted have also been defined.

Classification, then, operates solely with qualities. It uses intuitively discriminated qualities to create definable qualities at a higher level. No kind of quantitative information may be used in definition because units cannot be created utilizing continuous "attributes." To make use of quantitative information, it is necessary to convert it into qualities. The most frequently occurring example is the use of metric data such as size. Length, from which size is developed, is continuous because it is numerical. If, however, a set of things falls into two groups based upon length, i.e., length can be shown to be discontinuous, then the two groupings can be regarded as sizes (large and small). If, however, length were found to be continuous, no conversion into size is possible without an arbitrary decision external to the problem. Indeed, when things such as length are convertible into qualities such as size, they are generally perceived as qualities of size in the first place.

Quality and classification do articulate with quantity in two very important manners, description and distribution. This articulation follows from the basic assumptions about the ideational and the phenomenological realms and their articulation with each other. Classification, we noted, is useless without groups; groups, meaningless without classes. Phenomena are the ultimate focus of any inquiry, and groups are aggregates of phenomena. Groups must be described and cannot be defined. Classification provides a means of creating groups and a framework for distinguishing kinds of phenomena. The *denotata* of a class constitute a group. Identifying X items as members of Class A conveys only that the items all display the necessary and sufficient conditions for membership. The actual items themselves each consist of an infinite series of attributes, only a few of which are incorporated as features in the class definition. The other attributes which one cares to distinguish and which constitute the bulk of the items included as the *denotata* of a class are variable by definition. Some may be coterminus with the group; some, while restricted to the group, are not universal within it; and still'others which occur either in some or all of the items also occur in other groups. These variables, of course, can be spoken of as variable only after the framework of definitive criteria, itself invariable, has been established. As variables they can be controlled; this is description, and it is here that quantity enters. The description of any set of things can be accomplished only by means of quantitative statements. If one wishes to say what the *denotata* of a class look like, this requires a statement of variable features. It is usually done either by listing the variables or by citing their frequency of occurrence in the group. Usually a mean and a range for each variable attribute can be given. A description, then, is a quantitative generalization about a set of historical phenomena. As such it is bound to that set of phenomena. If a new instance is identified as belonging to the same class and thus is included in the group, the description of the group will change to accommodate the new case. One point of articulation between classification and quantification is that quantification of some kind is always required to describe the *denotata* of a class.

The second important articulation is in the realm of distribution. Classification enables one to identify only a given instance as a particular kind of thing displaying a definitive set of attributes. Identification is not a useful end product because it does not convey anything which was not already evident in the classification. A frequent course is to measure the occurrence of the *denotata* of a class in dimensions outside the classification, such as time and space. This can be done simply by plotting the occurrence of the *denotata*, producing maps or graphs analogous to simple lists of variable attributes in description. More sophisticated distributions are possible, and these require more complex quantitative information than simple occurrence.

Isopatch maps, for example, involve frequency of occurrence through space. Seriations, familiar to all prehistorians, are special kinds of distributions through time, again often based upon frequency of occurrence. The specific forms need not concern us here. If it is desirable to deal with distribution of *denotata* of a class in a given problem, this must always be done quantitatively.

Another important articulation lies in the matter of correlation. Rather than measuring the behavior of the occurrence of a given class against a constant dimension such as time or space, the distribution may be measured against the distribution of other similar units (covariance). The potential of these techniques is recognized in much of what prehistory does today, and most of the more sophisticated statistical operations are means of implementing this kind of inquiry. While less obvious, what a description describes, what a distribution or a correlation means, is a function of the definition of the units whose variable behavior has been measured. *Quantification articulates with classification in using, not defining, classes.*

In summary, classification is a process involving units, both as an input and as an output. Units are qualities, not quantities, and thus classification involves only qualities. Quantification necessarily must follow qualification and plays a role in employing the classes in given situations, but quantification cannot enter into classification itself.

3. CLASSIFICATION STATES RELATIONS ONLY WITHIN AND BETWEEN UNITS IN THE SAME SYSTEM. This third axiom of classification attends the nature of the informational "content" that is built into classification. Classifications are systems of classes and, as systems, are closed. Statements made about a system apply only within that system. The informational content of classification is thus completely internal. As has been indicated earlier, classification consists of a series of linked *significata*. It follows that these relationships are structural and that the content of a classification is entirely formal. Content in the sense of things and events is introduced by identifying the *denotata* of a class, but the class itself has no specific phenomenological content.

It further follows from the previous discussion of classification as a series of linked *significata* that two kinds of relationships must obtain in all classifications, relations within classes subsumed under the *significata*, and relations between classes subsumed under the links between *significata*. The first kind of relationship, those obtaining within classes, is universal for all kinds of classification. Chaos is ordered by stipulating finite kinds of things to which the infinite number of actual instances can be assigned. The *denotata* of a class are considered redundant; they are identical in terms of the criteria for membership in that class. The relationship that obtains within classes then is one of equivalence or identity. Indeed, this is the only reason for classification, to create redundancy. The notion of equivalence or identity needs to be further explored, for, given the assumptive basis of classification, identity must be a relative condition. Obviously, identity obtains only within a classification. The basic premise on which classification is founded assumes that no two things in the phenomenological world are the same. If they were, there would be no point to classification. Identity can mean only that *within the framework of a classification*, which is stipulated by the attributes chosen as criteria, things in a given set do not differ from one another. They obviously must differ from one another in respects not considered definitive. Furthermore, identity can obtain only when a scale is specified. This follows from the assumption that phenomena can be viewed at an infinite series of scales. What is unitary at one scale will be composite at a lower scale and component at a higher scale. Thus, not only is the notion of identity restricted to a definitive set of criteria, but also to a specific scale. The equivalences or identities embodied in a classification are the classes themselves. The *significatum* of each class is simply a statement of the terms of that equivalence.

The second kind of relationship that is embodied in all classifications is a relation between classes. The nature of this relationship differs from one kind of classification to another. However, there is one characteristic of between-class relationships that all kinds of classification share, namely, that this relationship is always an expression of some kind of non-equivalence. The non-equivalences which link classes in a classification are structured, and thus it is always possible to determine in what manner two given classes are non-equivalent. This is assessed by a comparison of the *significata*. For example, if Class 1 is defined by features *a-c* and Class 2 defined by features *b-d*, then the non-equivalence linking these two classes is identifiable as $a + d$. The specific forms of non-equivalence vary from one kind of classification to another, and this sort of variance will form the basis for the succeeding chapter on kinds of classification. Both the equivalences and non-equivalences embodied in classification have important bearing on the evaluation of classifications and will be treated in that context further along in the chapter. It is sufficient here to reiterate that: (1) classifications are formal; content is introduced only by identifying the *denotata* of a class; (2) two kinds of relationships obtain within all classifications, relationships of equivalence within each class and relationships of non-equivalence between classes; (3) classification provides a means of explicitly stating these relationships, the *significata* embodying the equivalences and the comparison of *significata* the non-equivalences; and (4) classifications, being formal structures, are organizing devices, not explanatory. Without content, explanation is not possible, and classification excludes all content from the start.

4. CLASSIFICATORY UNITS HAVE PRIMACY OVER LABELS APPLIED TO SUCH UNITS. This is not so much an axiom of classification as it is an answer to an issue. Obviously, classes must be identified by some device so that one can talk about them. They must be named, numbered, or otherwise provided with some kind of designation. Designation is an entirely arbitrary procedure outside of classification itself. Nonetheless a great deal of confusion often arises from a confounding of the label designating a class and the class itself. Semantic labels usually are inferences about a class (e.g., calling a particular kind of tool an axe, or biological species names part of an evolutionary scheme). A classification as a set of equivalences and non-equivalences enables one to say only whether a given thing is the same or different from another given thing. It cannot tell one why they are different, though how they are different is embodied in the non-equivalence. Why two things are assigned to different classes can be only a matter of inference and, as such, is outside the classification. A great deal of time has been expended on how classes should be labeled (for instance, numbers and/or letters versus words; if words, which words, etc.); however, the important thing to recognize is that the label can never bear a necessary relationship to the class. It is always a

label, simply a device to identify the class for purposes of discussion. What it is called is not important. The only necessity is that one be able to recognize the class by the label. A demonstration that a semantic label has been inappropriately chosen (e.g., showing that a class called axes are really hoes, or that species A. *pox* is really more closely related to B. *rash*) does not say anything about the classification, only about the inconsistency of some naming procedure.

5. CLASSIFICATION, CLASSIFICATIONS, AND CLASSIFICATORY UNITS HAVE PRIMACY OVER STRUCTURES, STRUCTURING, MODELS, AND MODEL-BUILDING. This is the least axiomatic, most commonsensical of the statements about classification. Models and structures are devices for illustrating relationships between classes which are not part of the same classification. It follows that one must first have the pieces before one can build something out of them and, furthermore, that the nature of the pieces is going to determine what kinds of things can be built.

Evaluation

Classification assumes that the phenomenological world is capable of order. To bring order and meaning to phenomena, four assumptions are made, two which locate the classes (level and field), and two which stipulate the means for ordering (distinguishing attributes and selecting some as definitive). The product is a set of equivalences (classes) and non-equivalences (relations between classes). Although obviously based upon observations of the phenomenological world, classifications are formal structures and lie wholly within the ideational realm. Lacking phenomenological content, they are not explanatory but, rather, organize and unitize the phenomenological realm so that it can be explained.

Explanation is apart from classification and based upon inferences about the organization that is imparted to phenomena by a classification and the distribution and correlation of phenomena so organized.

One final aspect of classification needs to be considered— evaluation. Irrespective of how the classification was formed, there are two elements involved in evaluation. Classifications are susceptible of evaluation as systems of the ideational realm in terms of their logical consistency. Further, they may be evaluated in terms of the choices exercised in the selection of the field, the scale of the classes, the discrimination of features, and the selection of a portion of these as criteria.

For a classification to be accepted as valid, it must be internally consistent. Decisions in the formulation of the classes incorporated in it must have been made with reference to a unified set of rules. Whimsical choices are not permissible for they destroy the system nature of the classification and negate any possibility of explicitly stating the relationships between classes. Examining a classification for internal consistency is an evaluation of the structure of the classification. If a classification is found to be inconsistent, it cannot serve as a classification because it does not provide any means of stating relations between classes.

The evaluation of classifications in terms of the four initial assumptions is much more complicated, because this is an evaluation of the classes themselves. The actual evaluation is of the choices made in: (1) selecting a field; (2) selecting a particular scale at which the classes are formed; (3) discriminating features for the creation of classes; and (4) selecting from among the discriminated features those which are to be considered definitive. Each of these sets of choices, if explicitly stated in the construction of the classification, is susceptible of evaluation in terms of parsimony and relevance.

To make such an assessment, it is necessary that the classification have a specific, explicitly stated purpose. Many, many "classifications" do not have explicitly stated purposes, and for this reason rational evaluation of the choices incorporated is impossible and the classifications have to be accepted or rejected on faith. Because this seems to be a problem area, it needs to be explored in detail. Perhaps the most frequently drawn, though implicit, criticism of purposeful classification is embodied in the notions of "universal," "discoverable," "natural," or "descriptive" classes. The assumption behind all of these notions is that there is some kind of order in the phenomenological world which is inherent, or more inherent than other orders, and thus what one has to do is discover the order rather than create it. This kind of assumption is, of course, counter to the assumptions upon which classification is based and appears to derive from a lack of understanding or concern, or both, of operations involved in the construction of classifications.

Be this the case or not, one important thing is decidedly clear. If one assumes that there are "natural" classes or "universal" classes, the products of classification are untestable, incapable of evaluation. Their evaluation necessarily rests upon the demonstration of the assumption of universal order, or inherent order, which obviously cannot be done. If, on the other hand, one assumes that one is creating an order, not discovering it, and thus must provide explicit statements about the choices involved rather than assuming that the choices are natural, these choices can be phrased as hypotheses about the relationship of the choices to the problem and to the stuff being ordered. We possess rational means of weighing the relative merits of hypotheses, and thus the assumptions which constitute the initial input into classification can be made problematical, testable, and a rational basis provided for using one over another. The utility of assuming only that phenomena are capable of order rather than ordered in some unknown but specific manner is obvious. The first assumption permits the possibility of evaluating the procedure; the second does not.

While the notion of "universal" classes is treated above as a kind of approach without purpose, it can also be looked upon as multipurpose classification, a classification which will serve for all problems. This view, too, requires careful examination. The desire for such classes seems to stem from a "need" to have a name for something, to be able to call a given thing the same thing forever and ever. Within our own cultural system this is, of course, possible. A cultural system is itself a filter which establishes by convention the relevance of certain kinds of criteria over others. The "universal" class would seem to be the application of one's uncontrolled common sense deriving from one's own cultural background to problems which lie outside that background. To create universal classes without assuming that some criteria are inherently more important for all purposes than others, and that there is a finite number of such criteria, requires that all attributes be considered definitive for the formulation of classes. Indeed, it is not too difficult to find this expressed in the literature as "all attributes have been considered" or something similar. This is clearly impossible. All attributes cannot be considered if the attributes are infinite.

But even more important is the nature of the product that would be produced if such could by magical means be accomplished. If all attributes were considered, the number of classes would equal the number of cases considered. There could be no difference between the classification and the phenomena themselves. There could be no kinds of things, and thus there would be no classification. This is certainly an unproductive view, for we already assume that everything is different from everything else. Elaborate procedures involving all attributes would provide nothing that one does not already know from the outset. The simple observation that such "classification" has never been done is ample evidence of its lack of utility. This particular approach has much in common with assuming that there is some kind of absolute "etic" level which lies beneath all other distinctions.

To establish a rational basis for evaluation of the selections that take place in creating classes, it is necessary that the reasons for those selections be known. The relevance of the particular choices made can then be weighed against the purpose of the classification. If a particular kind of organization is required for a given problem, the selections made can be weighed simply in terms of whether or not that organization has been achieved with those choices or whether a new set of choices is required.

Evaluation of the choices involved in classification does not end with weighing the classification against the purpose of the classification. It is quite possible that several discrepant classifications can accomplish the same organization. Some classifications will do so, however, in a simpler manner than others. Parsimony and elegance enter into the evaluation here. Some classifications use as attributes inferences about the material being considered (e.g., inferences about the function of tools, the manufacture of tools or parts of them, etc.). Classifications which make use of this sort of feature are not parsimonious when compared with those which use as features attributes of the objects or events involved. Indeed, the use of inferences about events or things as attributes can never be justified, for those inferences undoubtedly have a foundation in features of the events or things, and the features themselves can provide the identical organization as the inferences without involving the demonstration of the inferences.

Some classifications are more elegant than others. For

example, a given classification may produce many more discriminations or classes than are required for a problem. Another classification which produces those classes required for a problem and only those classes required is, in terms of elegance, the preferable classification. This latter condition, while it is the goal, is not frequently achieved, and so evaluation is really a matter of assessing how closely various alternative classifications approach it, rather than which one achieves such elegance.

In introducing each of the assumptive steps that must be taken to create a classification, the relevant sources for making the decisions have been indicated. The field and scale at which classes are established are usually controlled by the general theory of a discipline. These choices and the basis for making them, then, will be further considered in the specific treatment of prehistory in the second half of the book. The discrimination of features is obviously predicated on the establishment of the field and scale of the classes since the features must be discriminated at a scale beneath that of the classes. The choice of a particular set of attributes as criteria for classification and the number to be used (level) is predicated on the particular problem being considered and the kind of organization of phenomena required. Importantly, these selections must be susceptible to evaluation first in terms of their relevance (which requires a problem), and secondly in terms of their parsimony and elegance. The structure of the classification is evaluated in terms of its internal consistency. Further evaluation of the structure of classifications will be possible in terms of the use to which they are to be put after the various kinds of classifications possible have been considered.

Classification produces definable units which are capable of evaluation. The process does not differ structurally from common-sense, intuitive discriminations except that the process is explicit. Once the field of the classification is established, an analytic step is necessary to discriminate features to be used in creating units. The analytic step not only provides the means of definition by stipulating the conditions for membership in a given unit in terms of features but also provides the means of evaluation in its explicitness. Evaluation of a unit can be undertaken only when it is possible to assess the relevance of the defining criteria to the problem for which the classification is being created.

Classification

The role of classification in science is obvious. Classification is the means by which phenomena can be categorized and thus become subject to manipulation. It is not, however, the only means of categorization, but it does provide certain crucial elements not possible with other kinds of arrangement. The most important of these is the definable character of the categories. Since the categories can be explicitly defined, the means of identifying real phenomena can be accurately communicated from one person to another. Also, because the process of creating the classes is explicit, the units do not have to be taken for granted but are instead problematic, being subject to revision or change as demanded by evaluation.

Finally, the field of application is limited by the nature of classification. It can be used only to organize phenomena. It is entirely formal in structure and does not provide explanation, only organization. The organization may be used as the basis for inference, but this is a quantitative step beyond classification. Equally important, and closely connected to its organizational nature, classification must be problem-oriented. A single classification will not serve for all problems. The organization created by classification depends directly on the attributes treated as definitive of classes. The relevance of those attributes to a problem is the source of evaluation. Some organizations will be useful for some problems, but other organizations will be required for different ones. Except in the circumscribed context of our own social environments, a dog is not always a dog. He is a dog for some purposes only, and he is other things for other purposes.

3

KINDS OF
CLASSIFICATION

Introduction

*I*n the preceding chapter it was indicated that all classifications embody a statement of two kinds of relationships: relations within units which are always those of identity; and relations between units which serve to link classes together into a classification. It is this second category of relationships, those obtaining between classes, that determines the form of a classification and in turn results in kinds of classifications.

Apart from disparate usages of the term classification which effectively create different kinds of "classification," there is as much confusion evident in the literature, especially the non-archaeological literature, on kinds of classification as on any other aspect of the problem. Basically the confusion stems from treating classification as a single, unitary device, a failure to recognize differing kinds of relationships that can exist between sets of classes.

The tendency to treat all kinds of classification as essentially the same is particularly apparent in the natural sciences. This circumstance usually arises from the selection of one form of classification, perhaps on the basis of successful application, which is then traditionally employed to the exclusion of other forms. There is, of course, nothing intrinsically wrong with this procedure if, and only if, the problems investigated by that discipline are likewise unitary in nature and are of the same kind as the ones responsible for the initial selection. Unfortunately, this latter condition is not always the case. A brief example drawn from the biological sciences may serve to illustrate the point.

Since Darwin, the biological sciences as a whole have been preoccupied with the notion of evolution as the key concept in their theoretical structure for explanation. However, the classificatory devices employed, in particular the notion of species, in large measure antedate this explanatory concept. With the rise in importance of genetics in the biological sciences, the always vague notion of species has been made less vague by defining such units in terms of observed or stipulated genetic disjunctions, be they only regularly breeding populations or populations separated by actual breeding barriers. Importantly, however, the basis for defining species lies in disjunctions. Now, obviously, this notion of species is applicable to modern contemporary populations of animals. Logically, it is applicable to any set of contemporary animals be they in existence at present or at some specified and temporally restricted period in the past.

The hierarchic structure in which species were framed by Linnaeus and others had obvious similarities to the picture produced by the notion of evolution and the differentiation of species through time. Thus, in the nineteenth century, when investigators turned their attention to fossil remains, the notions of species and the hierarchic structure went with them—and in applying them to a new kind of problem, serious errors were committed. First, there are serious problems in taking any kind of unit like species and attempting to use it to organize fossil remains for explanation by means of the concept of evolution. Species must involve disjunctions, genetic or otherwise, to bound the units. But evolution assumes that all forms, similar and dissimilar, are linked, not by disjunctions in genetic material, but by continuities. The logical incongruity of the organizing concept and the explanatory concept is apparent; however, in initial practice it was not. The reasons are fairly simple. The fossil record is very incomplete. Real disjunctions in the *record* occur, though the development of which the remains are a record is continuous. Thus it was possible to assign a given set of fossils to a species without any great difficulty, because it could be separated from other related groups of fossils by gaps in the record (but not in genetic development). Once fossil lines began to be well represented by actual remains, problems began to appear, as the current state of man's own ancestry indicates. One is faced with arbitrary decisions as to whether a given fossil is to be placed in one or another species, solely because the fossil, in its form, lies between two previously created species, defined intuitively on the gap which the fossil in question now fills, thus the nonsense proposition that at some point in the evolutionary line an individual of one species gave birth to an individual of another species.

This example demonstrates other problems that are purely formal in character. The species notion was initially developed for application to whole animals to create an organization for whole animals. Genetics has expanded this to organization, not for individuals but for *populations* of whole animals. The fossil record unfortunately does not come in the form of whole animals, but pieces of their skeletal structure strongly biased by preservation characteristics in favor of skulls and teeth. For all practical purposes, fossil species are defined on the basis of skulls and teeth, yet the organization is assumed to be for whole animals. Obviously the species of the paleontological past and those of the modern world are not comparable. Further, unless one can posit a direct link between the form of the teeth and skulls of animals in general and the remainder of their bodies, paleontological species must be classes of skulls and teeth, not animals or populations of them.

Neither of these problems, both essentially functions of the relationships obtaining within classes, would have developed in the biological sciences were it not for the hierarchic structure in which the notion of species is embedded and which overtly parallels the notion of evolution. First, in the hierarchic structure only the species has a phenomenological referent; the units such as genus, family, etc., are entirely analytic units which serve to organize species and genera respectively and not real animals, their remains, or populations of either. Again the logical incongruity between the form of the classification selected and the assumed nature of the phenomena to be organized is evident. The analogy between the Linnean hierarchy and the differentiation of species through time paralleling the notion of evolution is thus misdrawn. Genera do not differentiate into species, but rather a species differentiates into several species.

Further, as will be considered in the body of the chapter, the particular form of classification chosen, irrespective of why it was selected, has inherent qualities rendering it something less than useful for the purposes to which "species" has been applied: namely, that no unit in a hierarchy, or a taxonomy as it is called, can be defined in terms of the phenomena being ordered, but only by inclusion in a higher level of the classification. Definition is by division, not by intersection. Initially this presented no problem to paleontologists because, as has been noted, the incompleteness of the fossil record furnished neatly separated groups which required only labeling. As the fossil record became more and more complete, the intuitive nature of species' definitions, indeed the real lack of definitions, became more and more obvious, and the suitability of "classical taxonomy" was questioned. Today controversy rages over this point in the biological sciences. New means of organizing fossil remains such as numerical taxonomy (not a kind of classification) have made their appearance in an attempt to correct the increasingly obvious inability of "classical taxonomy" to define species in anything but a mystical manner.

The movement to rectify these problems is not without serious errors as well. The protagonists of "numerical taxonomy" themselves often view "classical taxonomy" as the only kind of classification and, while still using the term classification, are attempting to introduce non-classificatory arrangement as a substitute, a device which is equally, though differently, ill-suited to the problem. The Linnean hierarchy, simply because it has traditionally been the sole form of classification employed, is taken to be the only possible form. Thus, inquiry into different, more appropriate forms of classification has been slowed.

This digression, of course, has been much simplified. It should, however, serve to demonstrate that important kinds of confusion greatly affecting the use of classification do exist in the sciences outside of prehistory. Much of the confusion focuses on the relationships between units, that is, the form of classification. These problems reflect a strong tendency to use classification as a technique rather than as a method. The assumptions upon which it is based are ignored when one learns "how to do it" instead of why it works, what it works on, and what the results mean. Failure to understand the assumptions has led to the application of kinds of classification to problems for which they are not suited. Because the assumptions are not made explicit, the conditions under which specific forms of classifications are applicable are not obvious and, further, no means of evaluating the results are possible. It is the contention in this chapter and, indeed, the volume as a whole, that much of the confusion results from the misapplication of a good method rather than the application of a poor one.

A Classification of Classifications

It should be obvious that to accomplish the aims of this chapter it is necessary to make use of the very device that is to be examined. Because such is the case, the classification used herein must be sufficiently explicit that it may be identified with one or another of the end products of this examination, and thus itself amenable to evaluation in the terms set forth herein.

In accordance with the discussion in the second chapter, the first step in any classification must be the definition of the field for that classification. In the present case this has already been accomplished, for classification has been defined earlier as the process of creating units of meaning by means of stipulating redundancy. Figure 3 shows its relationship to grouping. The field for the present classification can be taken to be classification as previously defined and as outlined in Figure 3.

In the same chapter it was noted that classifications consist of linked sets of *significata* or intensional definitions. Since the *significata* are the only tangible aspects of a classification, the second step in creating a classification, that of identifying the source of attributes, is relatively simple—the *significata* and their constituent elements are the only possible source. Obviously, some characteristics of *significata*, such as the nature of the constituent distinctive features, would organize classifications into classes based upon the kinds of classes contained within them. Our stated problem, however, is to examine the relationships between classes and the effect these relationships have on the form of classification. Thus, those characteristics of *significata* which are common or can be common to *significata* in general, and not those attending the content of individual classes, are relevant. There are many ways of looking at *significata* which demonstrate this kind of relationship. *Significata* may be differentiated in terms of the relations between constituent distinctive features (e.g., some elements in the definition may be more important than others, or they may be of equal weight). They might be differentiated on the basis of the processes involved in definition, the manner in which the *significata* come into being, or the way in which the distinctive features are associated with each other. Ultimately, then, in one form or another, selection of the elements constituting the *significata* is the characteristic useful for organizing classifications for an examination of the relations between classes and the effect this has upon the form of the classification.

In the right arm of Figure 3, which treats classification as of two kinds, these various ways of viewing the selection of features for class definition are summed up in the terms "internal" and "external." These labels derive from looking at the means by which the features are brought together into a *significatum* from the point of view of the objects included in a class. In one case the distinctive features can be associated directly from the objects considered; in the other case, the association of features is the result of a series of rankings within the classification at levels higher than the end-product classes.

Whether one considers *significata* in terms of their internal structuring or whether one considers various aspects of their construction processually, a quick conclusion is that in these terms *significata* are of two sorts: (a) *significata* whose constituent distinctive features are equivalent, unstructured, unweighted, and thus directly associated in analogous attributes of objects (intersection); and (b) *significata* whose constituent distinctive features are non-equivalent, structured, weighted, and thus inferentially associated (inclusion). Employing these two kinds of *significata* as criteria in a classification of classifications results in the recognition of two forms or classes of classification: one here called paradigmatic classification, employing the first type of *significata* (a); and taxonomic classification, employing the second kind (b). The following paragraphs will examine in more detail the characteristics of the two kinds of *significata* and the resulting forms of classifications.

PARADIGMATIC CLASSIFICATION: The concept "dimension" is useful for examining relationships between features in definitions, not only within the context of a single definition but also for classifications in their entireties. A dimension is a set of attributes or features which cannot, either logically or actually, co-occur. If there is one member of the set, then there cannot at the same time and place be any other member of the set. Further, all features belonging to a single dimension share the ability to combine with attributes not of that dimension. If A and B are members of the same dimension and I is a feature from another dimension, and, further, if AI occurs or is possible, then BI likewise must be possible. (Whether the combinations AI and BI actually do occur in the phenomenological realm is not important in assessing whether A and B belong to the same dimension, but, rather, only the possibility of their occurrence is relevant.) A *dimension*, then, *is a set of mutually exclusive alternative features*. Red and green are dimensional attributes. If something is red, it cannot be simultaneously green, but anything which is red could also be green. The dimension to which these features belong, of course, is the dimension of color, one which we ourselves use to categorize the phenomenological world.

Now, obviously, all features may be conceived of as dimensional in relation to other attributes, either as belonging to same or different dimensions; however, features may or may not be selected as criteria in a classification because they are dimensional. Dimensionality of the features included in class definitions is one of the important distinctions between the two kinds of *significata* and the resulting forms of classification indicated above. In the case of paradigmatic classification, each *significatum* consists of a set of features, each of which is overtly drawn from a different dimension. In the case of taxonomic classification, the set of features constituting a *significatum* may or may not derive from different dimensions since dimensionality is not considered in their formulation.

The differences become much more apparent when the classifications as a whole are considered. In paradigmatic classification all of the class definitions are drawn from the same set of dimensions of features. Individual classes are distinguished from one another by the unique product obtained in the combination, permutation, or intersection of features from the set of dimensions.

Figure 4 serves to illustrate paradigmatic classification by means of a simple case. Three dimensions are involved in the

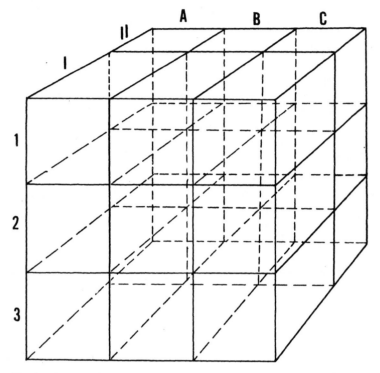

Figure 4. A three-dimensional representation of a paradigmatic classification of three dimensions (upper case letters, Roman numerals, and Arabic numerals).

classification: a dimension of Roman numerals, a dimension of Arabic numerals, and a dimension of upper-case letters. The first dimension consists of two features I and II; the second dimension of three features 1, 2, and 3; and the third dimension of three features A, B, and C. In each case it is assumed that the dimension is exhausted in the features, that is, that all possible representations of the dimension are covered by one of the features. The resulting 18 classes are simply the product of all possible combinations of these distinctive features, save that, by definition, features from the same dimension may not combine.

Dimensionality serves to control the possible definitive sets

A Classification of Classifications

of features. Individual class definitions will consist of one feature drawn from each dimension, the number of definitive features in each definition being a direct reflection of the number of dimensions used in the classification. The classification as a whole is united into a single system by the universal application of the dimensions. The features are definitive of the classes; the dimensions (as represented by the features) are definitive of the classification. *Paradigmatic classification*, when employed in this essay, is thus to be understood as *dimensional classification in which the classes are produced by intersection*.

Paradigmatic classes have some important characteristics which derive from definition by intersection of dimensional attributes. Firstly, all of the definitive criteria are equivalent; that is, none is or can be weighted over any other. In the example of Figure 4, Feature A is on a par with and cannot be included in Feature 1. The only weighting of attributes and dimensions that can be effectively accomplished is that of the selection of attributes and dimensions relevant to the problem for which the classification is intended (in Figure 4 the dimension of lower-case letters has been excluded and thus one might talk about the weighting of Roman numerals, Arabic numerals, and upper-case letters, as more "important" than other possible dimensions). This weighting, however, is done outside of the classification itself, and thus the choice of the particular dimensions employed can be phrased as an hypothesis, indeed must be so phrased, or completely ignored, and as such is amenable to testing, evaluation, acceptance, rejection, and revision. However, should it be deemed relevant to the problem attended by the classification in Figure 4 that the dimension of lower-case letters be considered, it would be added on a par with the other dimensions.

A second important characteristic of paradigmatic classes is that they are unambiguous, both in terms of their internal structure and in terms of their application as a means for creating groups of phenomena. This results from the dimensional characteristics of the features used in definition. All the features of a single dimension are mutually exclusive. Further, the combination or intersection of attributes to form definitions by dimensions prevents internal contradiction (e.g., that an object must be both green and red at the same time to satisfy membership conditions) from appearing in class definitions. From the standpoint of assigning phenomena to paradigmatic classes, the dimensionality of the defining features assures that, given adequate definitions of the features, each and every object or event for which the classification is relevant can be unambiguously assigned. X is either A or not A.

A third characteristic of paradigmatic classes is that they are comparable with all other classes in the same classification, and that the basis of comparability is explicitly established by the form of the classification. Paradigmatic classification, by virtue of being dimensional, considers only alternative manifestations of the same and specified dimensions. It is thus possible to characterize the relationships that obtain between classes in paradigmatic classifications as equivalent non-equivalences, that is, the structure of paradigmatic classification always specifies that all classes within it differ from one another in the same manner.

The field of a particular classification, of course, must be established prior to the formulation of the classification. In the case of paradigmatic classification, the field is often termed the *root* of the paradigm. The root is simply a statement of what the classes are classes of, and it is usually expressed as a feature or set of features common to all the classes within the paradigm. When this feature or set of features is added to the distinctive features which constitute the class definitions, it permits identification of the classification from which a particular class is drawn. It is important to remember, however, that the root or common feature in a class definition is not a product of the classification but is a symbolic record of one of the decisions made prior to the construction of the classification. All of the classes are defined within the classification. The root is not.

The number of dimensions employed in classification of this sort is determined by the problem for which it is being created. Obviously, the larger the number of dimensions and the larger the number of features in each dimension, the smaller the "space" covered within the field by each class. The number of classes will be increased. There is no limit beyond practicability to the number of dimensions and features within them that can be employed. In the case of features within a dimension, a dichotomous opposition (A and $\bar{A}$) is a minimal number. For graphic presentation such as used in Figure 4, the use of three dimensions is an obvious limit. However, simple listing of class definitions, or the use of graphic devices which do not use one dimension of space for each dimension of features, removes this apparent limit. As in the case of features, the minimal number of dimensions required is two, for without two dimensions intersection is not possible. It is, however, useful to consider as a special-case paradigmatic classification the *index*, treating it as *a paradigm with a single dimension of features*. The features in the dimension that constitutes the index are mutually exclusive, as is the case with other paradigms, and thus the classes formulated are unambiguous. The necessary and sufficient conditions for membership in such a class will be one in number; the number of features in a given definition is a reflection of the number of dimensions used in the classification. Since with but a single dimension classes are not formulated by means of intersection, indices are often treated as a separate kind of classification; however, because all of the differences between indices and paradigmatic classification relate to a single feature—the number of dimensions used—it is useful to think of indices as special-case paradigms.

In the practical business of formulating classifications this conception of the index is helpful. Each dimension of a paradigmatic classification is, in fact, an index, and such classifications are built up dimension by dimension. A major use of the index is the exploration of dimensions of features for paradigmatic classifications. Indices are capable of producing only simplistic orderings, and for this reason they are most commonly used for cataloguing and manipulating units (e.g., numerical and alphabetic orders) or for general problems requiring few classes (e.g., the classification for animals based on food-getting habits mentioned earlier, or the present classification of classifications based upon kinds of *significata*).

In employing paradigmatic classes to categorize things or events, identifying groups analogous to classes, the dimensional nature of the defining criteria is a definite asset. The necessary and sufficient conditions for membership registered as class definitions provides all that is required, and the only additional operation is the identification of features as attributes of objects or events. An event or object will be unambiguously assigned to one and only one class, or it will be found that the classification is irrelevant for the object or event (an expression of the fact that the instance lies outside the field of the classification).

Aside from the four sets of assumptions required of all classifications (scale, field, features, and criteria), paradigmatic classification, including the index, requires no further assumptive or inferential input. Paradigmatic classification is for this reason the most parsimonious kind of classification available, for, as will be shown, taxonomic classification requires additional assumptions. The use of paradigmatic classification requires only that there be a stated problem which in turn enables: (1) the definition of the field and the level at which organization is intended; and (2) the statement, in the form of a hypothesis, of the relevance of the definitive features to the problem. Once the relevance of the criteria to the problem has been stated, the classification is subject to evaluation through the hypotheses on which it is based. Most importantly, in the use of the units so produced, distributions and correlations have specifiable meanings. The investigator is not faced with a problem in which sets of units are found to bear certain relationships to one another but still lacking a means of stating the significance of the correlation or why they correlate. If the units are the product of a properly executed paradigmatic classification—i.e., all possible meanings that any correlations the units might have are known—they are overtly built into the units. The application of the units in a practical problem constitutes the testing of the hypotheses made in the classification. Unfortunately, far too little concern is given the formulation of classes. Thus classifications are rarely evaluated but rather become matters of convenience or opinion and the problem of what correlations and distributions mean must necessarily be treated as inference.

TAXONOMIC CLASSIFICATION. The familiar hierarchic structure of the taxonomy is, by implication from the preceding consideration, based upon non-dimensional distinctive features, at least as far as an entire taxonomy is concerned. (Portions of taxonomies may be considered dimensional.) A taxonomy is an ordered set of oppositions or contrasts which amounts to a division of the field of the classification into classes, sub-classes, and so on. Figure 5 illustrates the simplest form of taxonomy in which the contrasts are dichotomous oppositions. Classes, as defined units, may be formulated not only at the lowest level but at any or all intermediate nodes of opposition. The definition of any taxonomic class (taxon) is a record of the series of oppositions leading *from* the field to the class. From the point of view of any class, the definition derives from the inclusion of the class in a series of super-classes at higher and higher levels culminating in the field. As a result, the means by which the various elements or features in the definition of a taxon come to be associated (inclusion) contrasts with intersection in the case of paradigmatic classification. The features which make up the *significata* of individual taxons reflect the series of oppositions from field to class as a serial order, again contrasting with the unordered arrangement of features in paradigmatic definitions. The net effect of this serial ordering of the features of taxonomic defini-

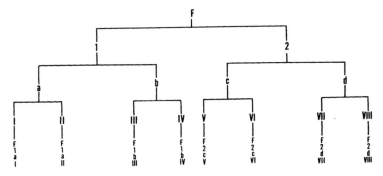

Figure 5. A taxonomy composed of binary oppositions. Only the definitions of the lowest level classes are written out.

tions is to restrict the range of the features constituting an opposition to a portion of the classification. In Figure 5, for example, the opposition d-b is relevant for the Superclass 1 on the left-hand side of the diagram. This does not mean that objects or events which might be assigned to VIII will not display attributes assignable to a or b, but that since they display Attribute 2, Features a and b will not be considered. This serial ordering of oppositions represents judgments as to the importance of the various sets of defining criteria. In Figure 5 the opposition between 1 and 2 is considered more important, more "basic" to the field, than the opposition between c and d or III and IV. Viewed again from the *significata* of individual classes, the various features that constitute a *significatum* are weighted from most important to least important. It is this weighting of features which is responsible for the serial ordering of features within *significata* and oppositions within the taxonomy. Ultimately, this weighting of features is the genesis of the hierarchic structure characteristically displayed by taxonomies.

It is not necessary, and in fact it is uncommon, that a taxonomy should display the symmetry of the example in Fig-

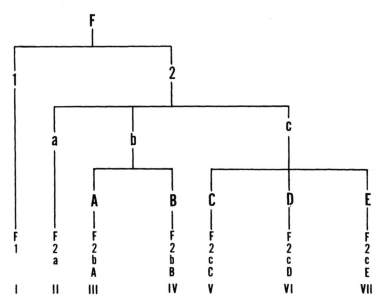

Figure 6. A typical taxonomy composed of various oppositions at different levels. Only the definitions of the lowest level classes are written out.

ure 5. Figure 6 presents a more realistic situation in which the series of oppositions leading to particular classes are not parallel in either number or kind. This diagram clearly illustrates the non-dimensional character of taxonomies and the restrictions placed upon subsequent oppositions by prior ones in defining taxons. In the case of Taxon I, a single feature serves to distinguish it from all other members of the field (I is, of course, redundant), whereas two features are required to differentiate Taxon II, three to distinguish III through VII. Ordinarily only the lowest level of classes need have empirical referents, that is, be designed to order phenomena, while the other taxons at higher levels serve to organize taxons at lower levels. An excellent example is the monotypic family as used in zoology in which the animals are categorized as members of a species rather than the family directly; the species in turn is the sole member of a genus which is the sole member of the family. This device is used to express a "degree of structural similarity" to other organisms in the Linnean hierarchy, here suggesting that members of the species in question are not closely related to other living organisms.

Taxonomy, then, is to be understood as *non-dimensional classification in which classes are defined by means of inclusion.* The relationships obtaining between classes are not uniform throughout a given classification. They differ from level to level (some classes include others) and also within each level. Thus the non-equivalent relationships which serve to separate classes are themselves non-equivalent and contrast with the equivalent non-equivalent relationships of the paradigm. There are additional characteristics of taxonomies which need to be considered, all of which derive directly from the defining characteristics noted above. It will be useful to examine these further aspects of taxonomies in conjunction with analogous features of paradigms where applicable.

Firstly, as a consequence of employing non-dimensional features for the definition of classes, the various distinctive features employed by a given taxonomy need not be mutually exclusive. Since the definition of a taxon involves not only a set of features, but also the serial ordering of those features based upon their "importance," it is quite possible (and not infrequent in practice) that distinctive features in one part of a taxonomy overlap features in another part. In Figure 5, for example, a and c can overlap each other without creating any ambiguities in the definitions so long as 1 and 2 are mutually exclusive. The opposition registered as a/b might represent a division of color into reds and blues with a encompassing everything from orangeish-yellows to reds and b encompassing the other end of the spectrum from greenish-yellows through violets. The opposition c/d might also register color, this time as violets and non-violets. Obviously there is substantial overlap in the coverage of a and d; however, insofar as the 1/2 distinction is made prior to the a/b/c/d distinctions there is no internal inconsistency. Further, as was touched upon earlier in the discussion, the a/b and c/d oppositions may represent different dimensions and thus may not be strictly comparable. In the above case, for example, the a/b distinction may represent colors while the c/d distinction represents textures. Any object which has color likewise has texture. If, however, the 1/2 distinction has been made prior to a/b/c/d distinctions, the former opposition will establish the relevance of one or the other of the lower-level distinctions and thus avoid any incongruence in the classification or ambiguity in assignments.

The non-dimensional character of taxonomies produces substantial potential for ambiguity in the assignment of objects or events. Taxons are unambiguous if, and only if, the serial order of the defining features is treated as a program for identification. The simple identification of a distinctive feature in a given instance is insufficient; the relevance of that feature is determined by all antecedent oppositions in the taxonomy. Perhaps the single greatest problem in utilizing taxonomies lies in this very thing. Unless the serial order of the defining features is stated, it is quite possible to make wrong assignments, or, worse yet, to be faced with an object which apparently belongs to two or more classes.

A second characteristic of taxonomies, one which also derives from the ordered nature of the defining features of the taxons, is that taxonomies have a non-permutable order. Since relationships between classes are not the same throughout a taxonomy, classes cannot be moved in relation to one another without altering the structure of the classification and necessitating changes in the definitions of other classes. Only the taxons arranged as members of the same superclass at the next highest level may be changed without changing the remainder of the classification. This contrasts with paradigmatic classifications which do not have any order in the defining criteria. There the classes may be changed in relation to one another without changing the classes or the structure of the classification. Figure 7 represents a three-dimensional paradigm displayed graphically so as to be comparable to a taxonomy and a comparison with Figure 5 clearly illustrates this difference. If the distinctions registered as 1 and 2 are exchanged for those registered as a and b there will be no resultant change in the

A Classification of Classifications

number of classes or in their definitions. The lowermost diagram represents a three-level taxonomy made up of dichotomous oppositions for the sake of simplicity. If the distinctions registered as 1 and 2 are exchanged for those registered as a and b, an entirely new classification will result. Neither the number nor the definitions of the new classes will be the same as in the initial classification. For this reason taxonomies are frequently referred to as non-arbitrary or natural in distinction to paradigms characterized as arbitrary and artificial. In this kind of discussion "arbitrary" is clearly being used in a sense

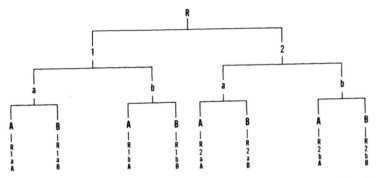

Figure 7. A three-dimensional paradigmatic classification displayed in two dimensions. Class definitions are written out at the bottom of the diagram.

different from that previously employed herein. It simply means that the position of any given taxon in the overall structure of taxonomy is fixed by the serial ordering of the defining criteria. *The position of a given class is non-arbitrary within the structure of the taxonomy;* the entire taxonomy, however, is arbitrary in the four respects that all classifications are arbitrary. Likewise, the feeling of "naturalness" that is imparted by taxonomies derives from the fixed order of taxons within the classification, for no classification is natural in the sense that the sets of equivalences and non-equivalences embodied in it are the only ones possible or even the best ones for all problems. While the non-permutable nature of taxonomies does not affect the pragmatic assignment of objects or events to taxons, it does tend to stifle evaluation of the classes and the classification; however, as is the case with the potential for ambiguity that inheres in taxonomies, intelligent use of this kind of classification, with any understanding of its limitations, can overcome the tendency for taxonomies to go without evaluation.

The third and final important characteristic of taxonomies is the assumptive or inferential input required in their construction. As has been pointed out, the serial order which is manifest in the overall structure of the taxonomy as a hierarchy involves ordering the oppositions by level and, by virtue of not being universal within the classification, some ordering in terms of positioning within a level. Only the initial opposition, the set of distinctions drawn at the highest, most general level, affects the entire classification. Subsequent ones are restricted to portions of the classification. For all sets of oppositions, an assumption of "importance" must be made to determine the order in which they are to occur. Further, for all but the initial opposition, an assumption of relevance must be made to position subsequent oppositions, those at lower levels. Since the various oppositions within a taxonomy are not dimensional, not mutually exclusive by definition, each specific opposition requires its own assumptions. The net effect is quite obvious. Taxonomies require a large number of assumptions as initial input for their construction in addition to the basic assumptions made by all classifications. In Figure 5, for example, 13 additional assumptions are required to determine the level and position of the seven oppositions. In larger, more realistic taxonomies the number of additional assumptions becomes proportionately larger. This situation is in direct contrast with paradigmatic classification which requires no further assumptions beyond those required of all classifications. Thus taxonomies cannot be considered parsimonious in relation to paradigms.

Given an alternative in the form of paradigmatic classification, it is reasonable to query how taxonomy is useful. If assumptions had to remain as assumptions, perhaps taxonomy would not be a useful device; however, if the assumptions are phrased as hypotheses which are testable and which upon testing have a high degree of probability, then the taxonomy becomes a much more parsimonious device. Unfortunately, this is not often done in practice and thus the intuitive qualities often ascribed to taxonomy. Indeed, this feature of taxonomy

A Classification of Classifications

lies at the root of the controversy between "classical taxonomists" and the "numerical taxonomists" in biological circles today. Given that taxonomy can be made more parsimonious than its structure initially suggests, it is useful to note the consequences of so doing. If the assumptions required by taxonomy must be phrased as tested hypotheses before taxonomy can be an effective alternative to paradigmatic classification, this means that, essentially, the outcome of the classification must be known beforehand.

If the classes must be known before a taxonomy can be constructed, serious limitations are placed on the utility of taxonomy. Taxonomy obviously cannot be employed to order a field of phenomena which is unknown in important respects. Further, of course, the assumptions must be capable of testing and positive verification, and this is not always possible even when a field is well known. Paradigmatic classification, on the other hand, is not faced with this problem because of its greater parsimony. For these reasons *legitimate usage of taxonomies is restricted to didactic purposes,* explaining in an elegant fashion a set of classes arrived at through some other means. Paradigmatic classification can then be regarded as appropriate for heuristic purposes, for the exploration and categorization of unknown or relatively unknown fields.

Taxonomy would be relegated to a minor role in scientific endeavor were it not for some advantages that it displays over paradigmatic classification. Firstly, it is a much more sophisticated device, capable of displaying more complex relationships between classes than paradigms. If a particular problem demands an organization of superclasses, classes, and sub-classes, paradigmatic classification cannot be employed, whereas taxonomy can. In fact, in any case in which non-equivalent relationships must be shown, taxonomy is the only classificatory system which can be used. The main advantage, however, is that taxonomies are far more elegant than paradigms. In the case of paradigms the dimensions of features are simply permuted for all possible combinations. Under practical circumstances this procedure will generate a larger number of classes than is required. Many classes may have no *denotata.* The delineation of those features which logically may be found in combination as opposed to those which actually combine with each other in the phenomenological world is certainly one of the major products of paradigmatic classification. However, for treating those classes which do have *denotata,* paradigms may be, and usually are, inefficient, creating a larger number of classes than required by the phenomena. A taxonomy, which restricts the combinations by ordering the oppositions of features, offers a way to generate those classes and only those classes which have *denotata.* The paradigm offers the means of determining what classes are required; the taxonomy provides the elegant means to arrive at definitions of those classes. However, without paradigmatically defined classes as a base, taxonomy remains an intuitive, unparsimonious device more often suspicious in character than not, and relatively useless without blind faith on the part of the user. Without paradigmatic classes as a starting point, the derivation of taxonomic definitions is a matter of faith, for there is no way to justify the choices made in its structuring.

Summary

There are two distinctly different kinds of classification which differ from one another in the relationships between classes and thus in the structure of the classification itself. In the first, paradigmatic classification, the classes are defined by means of unordered, unweighted, dimensional features; while in the second, taxonomic classification, classes are defined by serially ordered, weighted, non-dimensional features. The relationships between paradigmatic classes are equivalent non-equivalences. Thus all of the classes in a given paradigm are comparable with each other in a strict sense and, further, there is no inherent ordering among the classes, no fixed position which they bear to one another. Because no weighting, no internal judgments of "importance" are required by paradigmatic classifications, only the minimal number of assumptions required of all classifications are necessary. Thus paradigmatic classification is the most parsimonious form available, and it is particularly well·suited for heuristic uses, constructing initial classifications for given fields of phenomena. Further, since the assignment of objects to paradigmatic classes requires only the identification of attributes analogous to the distinctive features employed in the definition, this form of classification has the least potential for ambiguity in its application.

Taxonomic classification, on the other hand, stipulates

specific non-comparable relations among the included classes, producing the characteristic fixed hierarchic structure of the taxonomy. Since the features comprising the *significata* of the taxons must be weighted relative to one another, internal judgments of "importance" must be made to determine level within the structure and internal judgments of relevance must be made to determine position within level for all but the initial or highest level. Because of these judgments, the number of assumptions involved in taxonomic classification always, and usually greatly, exceeds the minimum number required of classification. Thus, taxonomic classification is the least parsimonious form of classification; however, this more sophisticated form of classification can embody more complex relationships than paradigmatic classification and provides an elegant form for generating a specific set of classes required for a problem or only those classes which have *denotata*. Taxonomy is legitimately limited to didactic applications where a solution reached through other means is to be presented in the most efficient manner. It cannot, by virtue of its lack of parsimony, be used initially to create a set of classes.

Criticisms currently leveled at classification are concerned almost invariably with taxonomic classification as outlined here. It has hopefully been shown that taxonomy can be a useful form of classification, though rather limited in terms of application. The reaction against taxonomy as employed in the evolutionary biological sciences stems from the misuse of the device and not from any flaw in the device itself. A common point of departure for such criticisms of "classification" (meaning taxonomy) is that it is subjective and intuitive. This aspect has been shown to derive from the large number of assumptions required to create levels and positions of oppositions within the hierarchic structure. The only possible means of making taxonomy more parsimonious is to be able to treat each of the assumptions as a demonstrated hypothesis, and this, of course, implies that the classes are already known from the outset. Without being based on prior paradigmatic classification, taxonomy is subjective, for the means of arriving at the classes is covert and untestable. In cases in which taxonomy has been so misapplied, it is likely that the investigator who has established the taxonomy had covertly employed paradigmatic classification to arrive at the set of classes embodied in the taxonomy.

Distinguishing between paradigmatic and taxonomic classification is then of considerable utility. Forms of classification which differ in terms of the assumptions required for their construction affect their range of applicability and the means by which they may be evaluated. This distinction between the unordered paradigmatic class and the serially ordered taxonomic class (taxon), and between the equivalent non-equivalences of the paradigm and the non-equivalent non-equivalences of the taxonomy, will be dealt with in the second part of this volume in examining the role, use, and misuse of classification in prehistory.

4

NON-CLASSIFICATORY ARRANGEMENT

*I*n preceding chapters that kind of arrangement called classification has been treated in some detail, for classification is the systematic foundation of science. As a result, this concluding chapter of the general consideration focusing on non-classificatory arrangement may seem out of place. The reasons for including a superficial consideration of non-classificatory arrangement are two: (1) the substantial confusion that exists between at least some forms of non-classificatory arrangement and classification, both paradigmatic and taxonomic, a confusion accompanied by an attempt to replace classification with one or another form of non-classificatory arrangement without a critical consideration of the consequences of so doing; and (2) as a corollary to this, the tendency to accept or reject classification or non-classificatory arrangement to the exclusion of the other and with little attempt to delineate the relationship between the two. It should be clear from the outset that non-classificatory arrangement, both in principle and as a technique, is not rejected here except as a substitute for classification in scientific inquiry. By the same token, classification must be rejected as a substitute for non-classificatory arrangement used in its proper role. The main aim of this consideration is to delimit the domain of both kinds of arrangement and to programmatically indicate the relations obtaining between the two in pragmatic terms.

Though in both archaeological and non-archaeological literature the kinds of arrangements grouped together here as non-classificatory arrangement are often labeled "classification," especially when there is an attempt to replace a classificatory scheme, all the forms treated here hold in common: (1) the absence of intensionally defined classes *as a product;* and (2) a concern with the phenomenological world in an at least overtly, theory-free context, resulting in the formulation of groups as end products. This fundamental difference between classification and non-classificatory arrangement was illustrated in Figure 3 where the latter is indicated under the heading of identification and grouping devices.

Since the differences between classification and the operations considered here are substantial, it is necessary to introduce two notions, those of group and those of similarity. The notion of group was touched upon in the introduction; however, expansion is crucial for a specific consideration of non-classificatory arrangement. *Group,* for the discussion herein, is to be understood as *an aggregate of actual events or objects, either physically or conceptually associated.* Groups are phenomenological—they have objective existence in their constitutent entities, although the "groupness," the association of the entities, is always in some measure non-objective. By virtue of objective existence, they are historical and contingency-bound. A group and each or any of its constituent entities exists at a given point in time in a given place. Groups have locations, not distributions, and so cannot be shared or held in common. As a result, the constraining boundaries of groups are not formal characteristics of the constituent entities, but rather are always ultimately reducible to temporal/spatial limits. Historical contingency is always incorporated in groups. When "definition" is used with reference to groups, one of two things is usually meant: (1) a statement of the temporal/spatial limits; or (2) an enumeration of the objects or events comprising the group or a statistical summary of same, that is, an extensional definition. An object or event cannot be assigned to a pre-existing group on the basis of its formal characters *without* altering the "definition" of the group. Being part of the phenomenological world, the construction of groups limits the data which can be considered to that finite set of cases incorporated in the original formulation. Groups always have a finite number of members in a finite time and space.

Non-Classificatory Arrangement

These enumerated characteristics of groups are readily recognizable as characteristics of objects/events in the common sense of the words, and all follow from the phenomenological nature of groups. Groups inhere in phenomena as aggregates of actual cases. The contrasts between groups and classes are obvious:

(1) Classes are intensionally defined on the basis of formal features of objects; groups are "defined" by enumerating and/or summarizing the members or by stating the temporal/spatial limits of the group.

(2) Classes are ideational units which exist independent of time and space and whose *denotata* can occur simultaneously at more than a single location or can occur at more than a single point in time at the same place, whereas groups are phenomenological and thus are governed by the physical laws concerning time/space/matter.

(3) As a corollary, classes have distributions; groups have locations.

(4) Classes are infinite in terms of their application, and any object or event acquired after the formulation of a classification can be assigned to a class without altering the definition; groups, on the other hand, are restricted to that set of objects/events originally incorporated in the group, and the addition of new information necessarily alters the "definition" of the entire group.

The consequences of these contrasts for pragmatic operations are a major portion of the basis for assessing the roles that classification and non-classificatory arrangement can legitimately play in scientific investigation.

In spite of these fundamental and seemingly apparent contrasts, certain kinds of confusion obtain in practice in differentiating classes and groups as a consequence of their analogous nature. The practical basis for this confusion lies in our own "common sense" environment. The assembled *denotata* of any class constitute a group in the sense used here. The problems in differentiating classes and groups stem from a confusion of the *denotata* of a class with the class itself. Where there is but a single classificatory scheme conceived possible, such as within a single cultural system or as within such sciences as are preoccupied with a single line of inquiry into a given subject matter, the pragmatic differences between attributes and features, between groups and classes, are negligible. In such conceptually-bounded circumstances the *denotata* of a class and the class itself are for practical purposes synonymous. Insofar as there are no alternative conceptions of a given set of phenomena, matters of definition, distribution, and application are trivial. Evaluation of the classes or groups is, however, impossible. When alternative classifications for the same set of phenomena are conceived possible or when evaluation of a set of categories is necessary, then the distinction between classes and groups, between the objects assigned to a class and the class itself, assumes paramount importance.

This consideration brings us to a most important point, the relationship of classes to groups, not in a formal sense such as outlined above, but in pragmatic terms. Classes are one means of associating the various constituents of a group. Assembling the extant *denotata* of a class, or a portion of the extant *denotata,* is one important means of creating groups. The necessary and sufficient conditions for class membership provide the means for creating the aggregate. However, the *denotata* of a class as a group consist of all the distinguished attributes of the objects/events included, not just the definitive features. Further, any set of assembled *denotata* is historical and contingency-bound. The actual *denotata* of a class viewed as a group are continually changing with the addition of new information. The assembled *denotata* of a class, while a group in the full sense of the term, are a very special case in which the criteria for creating the group remain contingency-free and thus capable of infinite expansion and incorporation of new information. The process of identification, the comparison of objects with the necessary and sufficient conditions for class membership in order to assign members, is the crucial link between classes and groups as represented by *denotata.*

The identification of objects with classes is not the only means of creating groups. A group can be created through any means of physically or conceptually associating objects or events. Groups can be created by arbitrarily drawing lots or by closing one's eyes and piling together things on one's desk. Most overt procedures for creating groups, however, make use of the notion of similarity, the second important concept in non-classificatory arrangement.

Unlike its counterpart in classification, identity, similarity is not precisely definable in a theoretical sense. In formal or phenetic terms, similarity is rephrased but not defined as "resemblance" of objects or events. In genetic (historical) or cladistic terms, similarity cannot be precisely defined in theoretical terms, for similarity is a relative state based upon the actual case being considered. Here lies an important contrast with the analogous notion of identity. Identity, too, is a relative state, but not relative to contingency-bound phenomena but rather relative to a given problem. Identity is determined in the context of problem, similarity in the context of phenomena. Similarity, then, is a contingency-bound notion which embodies a recognition of our earlier proposition that the phenomenological world is to be profitably conceived as an infinite series of uniquenesses. Identity denies the relevance of this proposition for a given line of investigation, and thus, being entirely within the ideational realm, permits demonstrative reasoning. Similarity, on the other hand, functions in the phenomenological realm permitting plausible reasoning.

Ultimately, similarity can be reduced to identity, identity of features of the objects or events being compared. The only means by which similarity can conceivably be defined or assessed is by the enumeration of *features held in common* by the compared instances. Such features, because of their recurrence from object to object, are obviously primitive classes. It is most unfortunate that this analytic classification is covert and intuitive in grouping procedures, especially since there is no *a priori* reason why it must be. If, however, the underlying analytic classification were explicit, grouping procedures would appear not as means of creating units but as means of stating the distribution of classes (features) over a given set of objects.

Be this as it may, two aspects of grouping must be emphasized: (1) Lacking a formal analytic step, groups cannot provide intensionally defined units which are capable of evaluation —the features upon which groups are based are assumed rather than treated as hypotheses with the resulting organization providing a test of the hypotheses as is the case with classification. And (2) because grouping counts and thus requires actual phenomena, the products are groups restricted in their organizing capacity to the data upon which they are based. The precision obtained with grouping devices is superficial, being a precision of mechanical manipulation rather than in meaning or utility.

Those devices which employ similarity as the central concept in group construction are generally polythetic, i.e., make use of a "large number" of features, or, more naïvely, "all features." The assumption lying behind this approach appears to be that there is but a single scheme for the delineation of features, so that number becomes a measure of "completeness." The notion of "all features" is, of course, contrary to our basic propositions about the phenomenological world, but, most importantly, it negates the basis for and the utility of the concept similarity, itself a means adapted specifically to deal with uniqueness and infinite (though not unpatterned) variability. For these reasons only the notion of "large number" requires any further attention. The necessity for "large numbers" of features derives from the relative nature of similarity in relation to phenomena. Given that similarity is a relative state, it must be assessed in degree rather than in absolute terms. Degree of similarity permits the "resemblance" of sets or pairs to be precisely compared and stated, and can be reckoned in many ways, usually in number of shared features or in percentage of shared features. Obviously the fineness of measure is a direct function of the number of features. The larger the number of features, the more discriminations of similarity that can be made, and the finer the measure of similarity, the more precision that can be achieved in creating and comparing groups. These similarity-based grouping devices aim at universally useful categories; however, as noted in Chapter 1, as the number of features considered is increased, the conceptual space covered by any combination of features is proportionately decreased so that the absolute number of categories increases. As the number of features approaches "completeness" the number of categories approaches the perceived number of phenomena, and the advantages of categorization in the first place are lost. Categories are reintroduced into similarity-based grouping by considering *degrees* of similarity. Groups may be formed by associating sets of things which share a certain number or a certain percentage of the total enumerated features. As shown in Figure 8, while most members of the same group constructed in this manner will share a majority of the same features, it is not necessary that any two things hold in common any features, for sharing can be accomplished through intermediate phenomena. This readily distinguishes classes and their *denotata* from groups. Groups do not necessarily have any constant, specifiable content analogous to the *significatum* of a class. It is for this reason that intensional definitions of similarity-based groups are impossible. The only means of definition is enumeration of the object or event included in the group.

The enumeration of shared features permits only the definition of similarity in a given case. It provides the terms in which similarity may be discussed; however, "definition" of the term necessarily varies from one case to the next. Because similarity by virtue of counts is contingency-bound, it cannot be

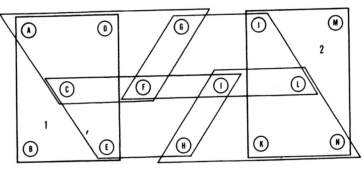

Figure 8. An extreme in group structure in which two objects (1 and 2) do not hold any features in common (letters in circles).

defined apart from each specific set of phenomena. Similarity, when used herein, is thus to be understood as *a quantitative assessment of the number of features shared by two or more objects or events.* Intuitive and non-quantitative assessments, while the basis of everyday similarity, are not usefully treated as similarity here, for their basis lies in a shared cultural background of the users and not in objective, overt statements.

As has been implied, within the category of non-classificatory arrangement it is grouping devices rather than identification devices which are seriously confused with classification, for groups are the phenomenological analogues of classes. Implicit, too, in the discussion of similarity and group concepts is that there are kinds of grouping devices as there are kinds of classification. Figure 9 presents a classification incorporating

Non-Classificatory Arrangement

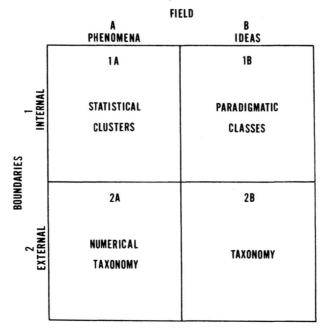

Figure 9. A paradigmatic classification of unit formation methods.

both classification and grouping, differentiating both in terms of the internal/external source of limiting parameters used in the discussion of classification. The analogous nature of classification and grouping and of kinds of classification and grouping are obvious. The internal/external contrast has already been explicated for classification. With reference to grouping, this contrast separates those kinds of grouping devices which create units by combination or association of features and which are herein termed statistical clustering, and those grouping devices which divide fields of phenomena by means of degrees of similarity, herein called numerical taxonomy. In the first case, paralleling paradigmatic classification, any set of groups is essentially equivalent and unordered, while in the second case, paralleling taxonomy, the sets of groups are essentially unequal and hierarchically ordered.

With the recent increased availability of computer time, experimentation with grouping demanding large numbers of calculations has resulted in a wide variety of techniques of grouping. For this reason, coupled with the fact that grouping as a whole is tangential to our main concern, the consideration of each form is a highly restricted sample, restricted with an eye to providing a background for a consideration of grouping as used in prehistory.

Statistical Clustering

The heading statistical clustering may be somewhat deceptive since all groups have characteristics that could be called "clusters," and, further, the term "cluster" or cluster analysis is not infrequently used with reference to some of the methods which would here be included under the rubric "numerical taxonomy." Statistical clustering is restricted to those methods which examine the *association* of attributes. A number of methods are available which make use of attributes (features in terms of the distinctions drawn herein), as the basic data input and which further create groups by summarizing the manner in which these features combine with each other in one or another kind of larger unit, usually discrete objects. All involve, overtly or covertly, some kind of "coefficient of association" and make only secondary, if any, use of the notion of similarity as the main device for the actual creation of groups. Techniques of this sort, while not as important in science in general as those termed "numerical taxonomy," have seen important use or at least proposed use in prehistory. Because of its simplicity which makes for good illustration and because it has figured prominently in prehistory's literature, chi-square clustering or sorting will be treated in some detail.

Like all of the techniques included as statistical clustering methods, chi-square clustering makes use of features as the initial input, features which must be mutually exclusive and dimensionally conceived. Ordinarily the operations involved are phrased as the *discovery* of consistently associated features, and thus the resultant groups are thought of as coherent bundles or clusters of features. Some methods simply calculate (record) the observed frequency of combination of these features and then examine these data for associations of high frequency relative to combinations and associations of low frequency or non-occurrence, that is, positive and negative coefficients of association. Chi-square clustering does essentially this, but additionally weighs the observed frequencies of combination of features in terms of the size of the sample being considered, taking into account sampling and the effect this has on association. Indeed, since grouping devices deal with phenomena, they all must take into account sampling before their results can be evaluated.

The basic procedures in chi-square clustering, *once one has the sets of dimensional features to be used and once one has a bounded, finite sample,* are fairly simple. First the frequency of occurrence of the features themselves is tabulated for all members of the sample. From this information can be calculated the expected frequency of combination. The expected frequency is obviously based upon the frequency of occurrence of the features alone and states how many examples of a given combination of features one would expect to find in the sample given its size. Expected frequencies are calculated for all possible combinations of features.

The second part of the procedure involves a tabulation of the actual observed combinations of features in the sample. The observed frequencies of combination or associations can then be compared with the expected number of occurrences. The expected frequencies represent the situation in which there are no tendencies for features to combine preferentially with others and thus represent random association.

The hypothesis made by chi-square clustering is that there are no patterned combinations in the sample. If the differences between each of the actual frequencies of association and the analogous expected frequencies are calculated in terms of units of standard deviation, the limits within which the observed frequencies can be considered a function of the sample can be read from tables and converted into statements of probability. Those frequencies which lie beyond the range of deviation attributable to the sample are then regarded as significant. If no frequencies occur which are significant, then the objects or events considered, in terms of the features used, are regarded as of the same kind. Both negative (frequencies significantly smaller than the expected number) and positive (frequencies significantly larger than the expected number) correlations may occur. In both cases special forces or rules are inferred to account for the non-random associations of features.

Non-randomness, then, is the *discovery* made by chi-square clustering. When put to the purposes of creating units, only the positive correlations are directly important, since the absence of a combination cannot serve as the basis of a group. The significant positive correlations are regarded as "natural grouping," and the objects which are not part of the significant clusters are treated as anomalous, fortuitous, or intermediate combinations of features. Further examination of the combinations of features using a covert notion of similarity can reduce the unaccounted for or anomalous combinations. On the basis of inspection those combinations which differ from the highly significant combinations in relatively few features may be grouped together with these latter, treating the less significant combinations as atypical or abnormal sub-groups or varieties. Ordinarily, a portion, sometimes substantial, of the original data remains unaccounted for as anomalous or intermediate occurrences.

The parallel of chi-square clustering with paradigmatic classification is apparent. Indeed, if viewed apart from its use in formulating groups of objects, chi-square clustering is nothing more than a statistical summary of the frequency of occurrence of the *denotata* of a set of paradigmatic classes. It is in its use as a means to create units that difficulties arise, first by delimiting units upon the frequency of occurrence of attribute (feature) combinations, which inextricably binds the units to a particular body of phenomena, and secondly by the use of similarity to further group units, which voids the possibility of intensional definition. Insofar as the frequency of association is used to delimit units, the units themselves are the product of happenstance—the product, for example, of which site happens to be known first.

While not structurally part of the method, the general attitude of "discovery" as opposed to construction of units contributes measurably to the difficulties, principally in discouraging the explicit statement of a problem whereby the features chosen can be tested for utility or at least justified. While the mechanics of unit formation are lucid and testable, their meaning is not. Thus not infrequently are the resulting units labeled "natural" or non-arbitrary. Aside from begging the question of meaning and utility, recourse to such labeling can usually be taken as a sure sign that the units have no specifiable meaning, much in the fashion that intuitive classifications are often called "descriptive." As for being non-arbitrary, it is difficult to imagine a device based upon paradigmatic classification to be any less arbitrary than that classification, if not even more so.

One further aspect of groups obtained by means of chi-square clustering which may not be initially apparent, but which is of fundamental importance, is the requirement of a bounded, finite sample. Since the group is an aggregate of phenomena, it must have temporal and spatial boundaries, if only past time and known space. Such boundaries are required for chi-square sorting for the coherence of the clusters. Insofar as "definition" of such a group is possible (either intensional definition where the classification is overt or enumerative definition where the classification is covert), the definition is in large measure a direct function of the boundaries of the sample and not its formal characteristics. If any new data are acquired, both the expected and the observed frequencies of combinations of features change accordingly, and with this the difference between the two calculated as units of standard deviation upon which the significance of the groups is based. Axiomatically the set of groups is restricted in application to the set of data which they comprise. The difficulties which obtain in attempting to employ such clusters for anything more than a statement of the observed distribution of *denotata* over a set of classes in a given case will be treated in some detail in the second half of the book. It should suffice here simply to point out that any confusion between groups obtained by chi-square clustering and classifications is one on paper, for the units are so widely and fundamentally different that if the units are actually employed any similarity disappears.

Numerical Taxonomy

While "statistical clustering" begins with features and formulates groups as associations or bundles of features that co-occur, the method here termed "numerical taxonomy" begins with the total set of phonemena to be grouped and in essence compares the constituent entities (Operational Taxonomic Units —OTU's) with each other formulating groups on the basis of similarity. In this respect there is an obvious parallel to taxonomy proper which begins with the field, analogous to the set of phenomena in numerical taxonomy, and divides and subdivides the field into classes. While numerical taxonomy, at least in primitive forms, has been employed in prehistory for thirty years or more, there is renewed interest in the application of the more explicit and sophisticated numerical taxonomy developed as an alternative to sloppy use of taxonomic classification in the biological sciences.

There are a number of methods, and it can be expected that the number will grow, given serious interest, which make use of similarity and which can be used to create units. For the purposes of illustration, numerical taxonomy making use of average linkage between operational units will be considered because it is currently the best candidate for application in prehistory as a means of creating groups, and is the simplest form of these similarity-based devices.

All of the similarity-based devices must begin by comparing in one manner or another all of the entities making up the set of phenomena to be grouped in terms of features. Similarity is assessed in terms of sharing of features between entities and expressed numerically as a coefficient of similarity. While some methods require one or another kind of coefficient, most are amenable to a variety of kinds. Thus the particular coefficient of similarity varies not only with the kind of device being used, but also with the ease with which it may be computed for a particular set of data or simply with a preference on the part of the investigators. The Brainerd-Robinson coefficient of agreement is perhaps the most familiar to archaeology. As noted in discussing the notion of similarity, the more features upon which an assessment of similarity is based the finer the discriminations possible. The practitioners of numerical taxonomy admonish the use of as many features as possible, not only to increase the fineness of discrimination but also to avoid "favoring" any one kind of characteristic—a pragmatically useful, but theoretically naive, proposition.

Coefficients of similarity are usually and conveniently expressed in a matrix in which each object or event is represented as a row and a column. The intersection of each row with each column is occupied with a coefficient expressing the similarity of the intersecting pair. The intersection of the row and the column representing the same object, of course, has the highest coefficient since it represents identity. There is an axis running diagonally through a matrix of these coefficients of similarity representing the comparison of each object with itself. All the information of a matrix is contained in half of the comparisons, on either side of the axis of identity, though for some purposes it is convenient to use the entire matrix. Numerical taxonomy is one of several methods of examining and reordering such matrices.

The basic procedures in numerical taxonomy begin with the inspection, either visually or with the aid of a computer, of the matrix for the highest coefficient not on the axis of identity and the pair is joined as a first-order cluster. The procedure is repeated until one of the units involved in a coefficient has already linked with another. Here a choice is presented. The first-order clusters may be linked directly on the basis of highest similarity of any one member with any one member of the other cluster. More common, though more complicated, is the average linkage method in which the mean of the similarity coefficients of all members of the previous cluster is computed and the new unit added only if this mean is higher than any other coefficient in the matrix. The procedure is continued, linking previously unlinked units in descending order of the coefficients or in terms of the highest average coefficient if previously incorporated in a cluster, until the entire set of phenomena is linked into a single cluster. The series of linkages can be conveniently recorded in "dendrograms" resembling taxonomic hierarchies (Figure 10). Beginning with any branch, the history of linkages with other units can be traced through the last linkage, uniting all of the units into a single cluster. At this point, however, there is only the total set of phenomena which were to be grouped and the constituent members of the group—which is just what you started with. The record of linkages made, however, provides the basis for segregating groups, and, since linkages are made in serial fashion, segregation is always potentially hierarchic. Groups may be created by vertical divisions based on the coefficients of similarity so that clusters with linkages above X value can be considered first-order groups; those above Y but less than X, second-order groups; and so producing a series of groups bearing a superficial resemblance to a taxonomic classification. Of course, unless some specific relationship between a given value for a coefficient of similarity and a problem can be demonstrated, such grouping is entirely arbitrary in the common sense of the word.

Another means of formulating groups is to inspect the dendrogram for disjunctions in degree of similarity and divide groups at these disjunctions irrespective of a given absolute value for the coefficients of similarity. Breaking the large cluster into groups which are internally quite similar relative to other

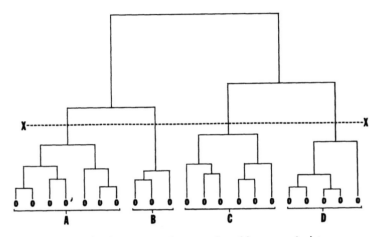

Figure 10. A dendrogram such as produced by numerical taxonomy. The four groups (A-D) are a product of dividing in accord with a given level similarity (X). The vertical distances of the lines connecting OTU's (0) represent the degree of similarity.

such groups may initially seem to produce "natural" groups, and in a sense they are "natural," but only so within the finite set of phenomena grouped. Irrespective of the means chosen to formulate the actual groups, the set of groups is bound to the given set of phenomena. Any additional data will change the composition of the groups, may alter the order of the linkages, and, particularly in the second kind of group formulation, change the entire pattern of groups. Definition presents serious problems too. Intensional definition is impossible for the members of any group may or may not have a common set of distinctive features. Indeed, as indicated in Figure 8, they may have no common features. Hierarchic structuring of numerical taxonomic groups is a function of choosing to break groups by level of similarity. Those formulated by breaking clusters at disjunctions are not necessarily hierarchic in relation to each other. The meaning of the groups obtained in either fashion is problematic. Without an overt classificatory basis, there is no means of assessing what similarity or resemblance means in a given case, whether or not it has been assessed in terms relevant for a given problem. With the tendency to polythetic treatment of features, alternative means of assessing similarity beyond using different kinds of coefficients are not usually considered; yet obviously, if the coefficients were based on an entirely or only partially different set of features, the entire structure of linkages in terms of the coefficients would be different, and thus the groups different. As in the case of statistical clustering, even enumerative definitions of the groups resulting from numerical taxonomy are in part a direct function of the boundaries of the set of phenomena grouped and not their formal characteristics. If at random half the units grouped were removed, the linkages would change; or if the number of entities treated were doubled, the linkages would change, and any change in the linkages would produce an altered set of groups. Thus, like statistical clustering, serious limitations are placed on numerical taxonomy as a means of creating units simply because the end-products have the characterictics of groups. By virtue of employing the notion of similarity, numerical taxonomy has further limitations not necessarily imposed upon statistical clustering. Statistical clustering has a basis in paradigmatic classification, and, when this is overt, clusters can be given meaning; numerical taxonomy lacks a classificatory basis, having only a covert analytic step resulting in the features used in assessing similarity, and thus cannot be given any meaning beyond the rather vacuous label "natural."

Identification Devices

This kind of non-classificatory arrangement can be conveniently separated from other kinds of arrangement in that identification devices are not a means of formulating units. As such they lie outside the general classification presented in Figure 9. Identification devices are of concern here only in that:

(1) they superficially resemble both classificatory and grouping systems because they are comprised by ordered units and thus are easily confused with these kinds of arrangements; and (2) they constitute a major means of actually employing other kinds of arrangement.

The aim of an identification device is the assignment of events or objects to categories that are established through some other means. Given that groups are restricted in application to the data from which they are derived, it follows that identification is a notion applicable only to the articulation of classification and phenomena. The term *identification device* is thus to be understood as *any formal structure designed to assign events or objects to previously defined classes.* Bridging the distinction between the ideational and the phenomenological, identification devices are highly variable in many of their formal characteristics. They must be adapted to specific data and classifications that they serve to link. Nonetheless, all have more or less the form commonly called keys. Of importance here is the general nature of such devices and the role they play, enabling one to distinguish them from the formally similar unit-formulating arrangements. This is perhaps best done by examining the construction of a key for a paradigmatic classification.

Figure 11 shows a hypothetical paradigm of three dimensions each comprised of three features with a root, I, a permutation of which yields 27 classes, each of which is denoted in the diagram by its definition written at the right. Many more classes than actually have *denotata* are generated; in this case only 11 classes have *denotata*, those marked with boxes to the right. If a new object were to be assigned to this classification, its features would have to be compared until it was matched with an identical definition. This is obviously inefficient. The key presented in Figure 12 represents a more efficient way to locate the appropriate classes for a given object. Through an ordered set of binary oppositions, those specific classes which have *denotata* can be quickly located. By examining the new object for each feature in the order in which those features occur in the key, the unprofitable comparison of the objects with inappropriate classes is avoided, and the investigator is led directly to the proper assignment. The ordered set of oppositions is simply a summary of what is known about the occurrence of *denotata* with respect to distinctive features and excludes all information

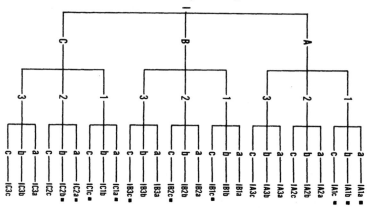

Figure 11. A paradigmatic classification of three dimensions with three features comprising each dimension. The boxes to the right of the class definitions indicate those classes which have **denotata** for the example in the text.

in the paradigm not relevant to the assignment of objects. By means of binary opposition, keys can facilitate the identification of objects with taxonomic classes even though these are more elegant than paradigms. The utility of identification devices increases with the complexity of a given classification and the number of possible class assignments. It is particularly useful for paradigms which generate a much larger number of classes than actually have *denotata*. Obviously in those cases in which the classification is simple and the number of possible assignments small, the time and effort involved in constructing an identification device is not justifiable, for there will be little appreciable increase in efficiency of identification.

Since binary oppositions are employed, the key can be considered dimensional; however, the number of dimensions (equivalent to the number of oppositions) bears no direct relationship to those of the parent classification, nor need the features within a given dimension be the same. For example, in Figure 12, if a given object displays Feature 1 it is necessary only to distinguish Feature a rather than a, b, and c. If a is the quality of opaqueness, b translucence, and c transparency, it is necessary only to ascertain whether or not an object displaying Feature 1 is opaque

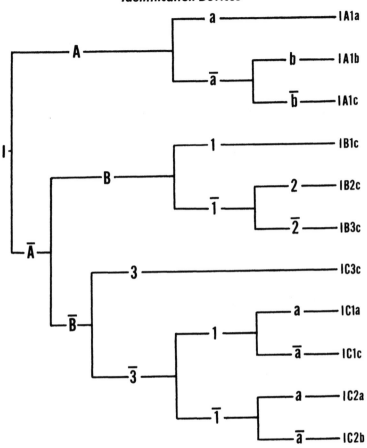

Figure 12. A key for those classes with **denotata** in Figure 11.

to assign it to the proper class. Without the key, all new objects would have to be segregated into opaque, translucent, and transparent, for in *some* cases all distinctions are required for proper assignment.

When diagrammed, identification devices resemble classifications; however, this similarity is superficial. Identification devices provide only a series of steps to rapidly identify an object. The classes are not defined within it. A comparison of Figures 11 and 12 will show that the class definitions cannot be derived from the oppositions used in the key. Further, there is no universal set of distinctive features. As is the case with taxonomies, the parsimony of the paradigm has been lost in achieving a more elegant system. In contrast with taxonomy, keys, such as portrayed in the example, do not necessarily have fixed, non-permutable order. In the hypothetical key, the last two oppositions in the branch leading to classes 1C2a and 1C2b could just as easily be reversed with no alteration made in the identifications or in the efficiency of the scheme. Whether or not the order in a given key is fixed is not a function of the key itself but of the definitions of the classes for which it provides identification.

While identification devices are not bound to any specific set of phenomena by virtue of their ideational element, they are restricted to phenomena assignable to classes previously known to have *denotata*. If an object assignable to a paradigmatic class not represented in the key were to be identified using the key, it would be given an assignment not in agreement with the class definition. Because such an object is given an unambiguous assignment such misidentification easily escapes detection. For this reason, identification devices are best used upon kinds of data which are well known. If the oppositions used in the key are not of the A/Ā kind, that is, mutually exclusive and exhaustive, misidentification will be replaced by ambiguity or no identification, which in turn permits detection of the new class member.

Keys or identifying devices can be constructed for any kind of classification and are restricted to classification; however, unless so doing increases the efficiency of identification over comparison with class definitions, there is little point to their construction. They are not a necessary adjunct to classification, though they can be exceedingly useful in applying complicated classifications. Undoubtedly the greatest problem found in scientific literature involving the use of keys is the substitution of a key for a classification. When the classification on which a key is based is not made explicit, either in the mind of the investigator or the work in which it is employed, it is difficult because of their similarity to detect which has been used, a problem especially difficult in written sources. A great deal of misunderstanding and an inability to replicate other workers' results can be a consequence.

Summary

To summarize non-classificatory arrangement it is well to look at the relationships that this kind of arrangement bears to classification. In many formal aspects all arrangements are similar. All involve units or categories of one sort or another. All provide some kind of structuring between such units, and the structuring, together with the units, appears as a system. Similar graphic devices as well as similar language can be used to present and manipulate all kinds of arrangements. Thus it is a simple matter to confuse one kind of arrangement with another, especially if the explication is less than complete. A comparison of the various kinds of arrangement shows that the consequences of such confusion can be serious indeed, and in many cases difficult to detect.

The initial problem faced by the student is the identification of the various kinds of arrangement as they are expressed, often covertly, in the literature. Classification can be distinguished by: (1) the lack of objective existence of the units, giving them an ahistorical character and permitting the simultaneous occurrence or sharing of their *denotata* recognizable as distributions—only classes can have distributions; and (2) the ability to provide intensional definitions for the units. Not a feature to be found in every case, but one of utility in recognizing classifications, is the presence of specifiable problems to which they are directed. Grouping devices may be readily identified by: (1) the fact that the units always consist of aggregates of objects or events with locations in time and space; (2) the inability of the units to include additional members without redefinition; and (3) "definitions" which derive from the historical boundaries of the sample used in the original formulation and which take the form of enumeration or summary of the content of the units. Identification devices are easily distinguished in that they have neither members nor definitions. Classes are all form, groups are all content, and identification devices have neither (or both, as you care to view it).

These clues to the various kinds of arrangement only partially permit identification, for when investigators have not maintained such distinctions in their work they often will shift from one form to another as a matter of convenience. This is certainly true, as will become apparent, for prehistory and its literature, and it means that each work must be carefully examined for consistency in matters of arrangement.

Since the focus of our concern is the creation of units and not their manipulation and use, identification devices can be disposed of quickly, for they do not formulate units nor is it likely that they can be confused with unit formulation. Their relationship to classification is simple and direct. They function to aid the identification of new objects or events with previously established classes. They are useful when and only when: (1) the classification is large, and there are many possible assignments; (2) the number of classes without *denotata* is large; and (3) the body of phenomena being identified is well known in its general characteristics.

The relationship of grouping devices to classification is likewise simple but not nearly so obvious. Class *denotata* once assembled by identification always constitute a group; however, this is quite different from grouping objects or events to construct units. Statistical clustering has a fairly patent basis in paradigmatic classification, and, the claims of its users notwithstanding, it simply *selects* some of the paradigmatic classes as important in a given historical case. Further, less important classes (numerically) may be merged with the important ones utilizing covertly the notion of similarity. Both the selection and merging are based upon counts deriving from a particular set of data, and it is just this feature which limits the utility of statistical clustering. Groups so produced cannot be defined save by drawing a line around them—they are what they are simply because they are. Such clusters have locations in time and space and cannot be used to measure variation in either dimension. To attempt to employ such units in an examination of variation is not terribly unlike measuring with a rubber yardstick of continually varying calibration. These criticisms apply *only* to statistical clustering as a means of creating units. If the underlying classification is explicit, these same procedures result in a statement of the frequency of occurrence of the *denotata* of a classification in a given historical case, evaluated in terms of the sample size. Such procedures have demonstrable utility in manipulating classes and in formulating and testing inferences about their behavior. The kind of grouping device termed "statistical clustering" cannot be regarded as a legitimate means of unit formulation, but it is a highly useful means of manipulating previously formulated classificatory units.

The relationship of numerical taxonomy to classification is less obvious. The notion of similarity is the basic point of difference, for there is no question that numerical taxonomy can formulate units. As with other kinds of grouping, however, there is no way to discover, at least programmatically, what the groups mean in relation to a problem. The units so produced as groups are subject to all the criticisms voiced of statistical clustering. The meaning, the kinds of inferences which may be based upon such units, is problematic and intuitive (thus the labels "natural" or "descriptive"). The situation which has given rise to the development of numerical taxonomy, the abuse of taxonomic classification, is certainly in need of correction; however, it is difficult to see how a similarity-based group of bones is any improvement. The context of its recent development does provide a key to its relationship with classification. If treated not as a means of creating groups, but as a means of treating the *denotata* of pre-existing groups, a useful relationship with classification can be stipulated. In this case numerical taxonomy summarizes' the occurrence of both distinctive and non-distinctive features over the *denotata* of a classification. As in the case of statistical clustering, numerical taxonomy provides a valuable means of manipulating class *denotata* and formulating and testing inferences about their behavior. With an overt classification establishing the groups, the arbitrariness in breaking groups at levels of similarity or using disjunctions in the sample is eliminated. The notion of similarity functions quite adequately in the realm of finite historical data but cannot serve as a means of creating units to frame ahistorical laws governing the behavior of phenomena.

Classification remains the only legitimate means of constructing units for scientific purposes. The procedures used in grouping devices, while not useful for the construction of units, are useful in the manipulation of class *denotata*. Identification provides the means of creating groups of utility. The development of grouping devices as *substitutes* for classification is a function of the misuse and poor explication of classificatory systems.

With a few notable exceptions, the new archaeology employs the methods described here as grouping devices in their proper roles as means of stating and correlating the occurrence and distribution of *denotata* of otherwise defined classes. Archaeology's predilection for borrowing from other disciplines does not bode well in this respect, for particularly in the biological sciences have grouping devices gained some currency. The temptation to employ these mechanically lucid devices to create units is deceptively enhanced by their explicit accounts of what is already known. Where they fail is less obvious; they cannot be employed heuristically, and they are not testable in any meaningful sense beyond their mechanics.

part 11

SYSTEMATICS IN PREHISTORY

5

PREHISTORY

As ought to be apparent at this juncture, the considerations undertaken as Part I of this volume do not constitute a fully developed, coherent theoretical system. Not only is such available elsewhere in various forms, but a treatment of this nature would far exceed the requirements for introducing some coherence into the formal operations and units of prehistory. Our goal is a coherent theoretical system for the formal aspects of prehistory, a much more limited goal than a general system. Thus the general considerations have been restricted to the explication, in terms adapted to current prehistory, of key notions—key in the sense that they are or ought to be issues within prehistory.

The initial problem to be faced in constructing a theoretical system for prehistory in formal aspects, and the one attended in this chapter, is that of defining the field for consideration. There are, of course, many possible ways of accomplishing this, and the choice of means as well as the end result have important consequences for all further operations and require deliberation. Indeed, one source of confusion in the literature of prehistory has been attributed to the vague notion of what prehistory is and what it is or ought to be doing. This vagueness undoubtedly reflects the unstructured manner in which prehistory has developed primarily from Old World antiquarianism—sometimes in conjunction with the natural sciences; sometimes, as in this country, in close conjunction with sociocultural anthropology;

and sometimes, at least effectively, in isolation. As long ago as 1953, it was possible for an eminent American prehistorian, A. C. Spaulding, to summarize prehistory as being that which prehistorians like to do—and nothing or little more. In many if not most quarters today this characterization is still accurate, the only important difference being that some prehistorians like to do things which their colleagues of twenty years ago had not thought of doing. There is, of course, nothing wrong with prehistorians' enjoying what they do; this is healthy, a requirement of a viable discipline. However, serious difficulties arise when this kind of characterization is the only accurate means of defining prehistory.

Prehistory has been defined many times and in various ways, this fact itself contributing in no small measure to the vagueness surrounding the meaning. Universal acceptance has not been accorded any definition, at least in part because all the definitions are more or less substantive, tied to a given area or problem. Boundaries around the field are drawn in terms of time and space (e.g., using the literal meaning of the label prehistory), or definition is in terms of specific goals such as "cultural reconstruction." Even if one or another definition of these sorts should gain currency, the vagueness attending the field of prehistory would have been merely shuffled under the academic rug. Insofar as the definitions are substantive, either definitions in terms of subject matter limited in time and space or definitions in terms of results, they do not specify how the field operates. One can do anything with a given subject matter, yet not all treatments of the preliterate past would be considered prehistory even by those who employ a history–prehistory distinction in the definition of the field. Likewise, not all "reconstructions" of the past would be admitted as prehistory, especially those frankly based on speculation, by those workers "defining" prehistory as reconstruction.

As was pointed out on several occasions in earlier chapters, neither the subject matter nor even the results serve to adequately separate the various academic disciplines. Rather it is theory, the manner in which a particular discipline views phenomena, that distinguishes the various disciplines and sciences. A particular view of the world will always be more relevant to some kinds of things than others, a feature which lies at the root of the subject matter approach to definition. Likewise, a particular view conditions the kinds of results possible. Especially with today's trend toward multidisciplinary study, the relevance of a particular discipline to a particular subject matter is continually expanding. There is, even in the space of a few years, ample demonstration of the independence of discipline and subject matter. Likewise, unanticipated results are not infrequently obtained, sometimes completely reorienting disciplines in terms of the "thing to do." These arguments do not mean that subject matter is unimportant for the definition of a discipline, but rather that the form of the subject matter, the way in which it is conceived, and not what it is, must be used, and, further, that subject matter alone or in combination with results is insufficient. If prehistory is to be an academic discipline and a science it must be *a kind of study*, not solely *the* study of a kind of thing.

One way in which prehistory could be defined consistently with the above discussion is to first develop a formal theory of prehistory and then define the field as that in which this theory is operative. While consistent and certainly accurate, this circular approach does not convey much information and would only contribute to the vagueness surrounding the meaning of prehistory instead of providing a basis for departure. The definition to be presented here, along with the explication of the terms used in it, is hopefully informative while at the same time consistent with the requirements of such a definition.

Given that prehistory has grown like Topsy, any definition, save that prehistory is what people who call themselves prehistorians do, is bound to exclude some things done under that label and perhaps include others not usually conceived of as prehistory. The definition to be presented has the advantage of including much of what is done under the label, and, further, the substantive definitions can be viewed as special cases of adaptation to a specific area, specific data, or a specific problem which holds the interest of a given investigator. Substantive definitions are not "wrong," but they are limited to the problem or data they are designed to serve. A general definition provides not only a means of discussing prehistory in theoretical terms but at the same time it provides a means of deriving the substantive definitions and enables one to link these definitions to one another rather than treating them as competitive, contradictory, and inconsistent.

In spite of an attempt to define prehistory in such a manner as to include much of what is done under this label, some kinds of activity and some specific studies are, of course, excluded. These exclusions result primarily from a failure of the activity or study to meet the requirements of science rather than on other grounds. Their exclusion here does not mean that they are not worthwhile, profitable, interesting, or entertaining. It simply means that they are different in important dimensions from the other activities considered and cannot be judged by the same yardstick. It is not asserted, for example, that "amateur archaeology" is not worthwhile or that theologically-based speculation on man's past is not interesting. They cannot, however, be evaluated by the same means as used here.

With these points considered, *prehistory is* defined herein as *the science of artifacts and relations between artifacts conducted in terms of the concept culture.* This definition stipulates: (1) the kind of study—science; (2) the main concept with which explanation is undertaken—culture; and (3) the manner in which phenomena must be conceived—artifact. Insofar as any given work conceives its data as artifacts and uses scientific means to achieve explanations framed in cultural terms, it is prehistory and within the realm of our examination. The remaining portions of this chapter will first explicate each of the three key notions involved in the definition and then examine the implications of this definition of prehistory for the relationships to other sciences and non-sciences closely linked to prehistory.

Science

In view of earlier discussion there is little need to further belabor this notion. Insofar as a given discipline has a theoretical structure which is employed to systematically organize phenomena for the purposes of explanation of these phenomena in a manner capable of testing, it may be considered a science. Employing this criterion excludes: (a) intuitive, non-rigorous approaches by virtue of a lack of overt theory and testability; (b) approaches which focus upon ideas rather than upon phenomena (e.g., philosophy); and (3) "descriptive" approaches which do not have explanation in the sense of prediction and/or control as an end product or a possible end product. A casual survey of literature bearing the label prehistory might suggest that it generally fails to meet these criteria, particularly in a lack of theoretical structure and testable conclusions. It is the contention here that this impression is more apparent than real, that at least as far as formal theory, systematics, is concerned, most of what has already been done in prehistory meets this criterion, but implicitly rather than explicitly. Further, while most of its conclusions are untested, they are testable.

Artifact

Unfortunately, there is no generally accepted definition of the subject matter of prehistory, again because of the substantive preoccupation of the discipline. The many definitions in the literature reflect the requirements of particular problems, kinds of problems, and areas, and thus are not suited for theoretical use or, for that matter, practical application beyond the particular problem or area for which they are developed. This lack of unity has been customarily dealt with by ignoring it— apparently no thought accorded to the non-comparability and contradiction that such fundamental disagreement introduces into the product of different investigators' work.

The concept artifact must be treated as a kind of theoretical template which segregates those phenomena of interest and amenable to scientific study by means of the concept culture and thus imposes a particular view upon the phenomena so segregated. The term *artifact* will herein be understood to mean *anything which exhibits any physical attributes that can be assumed to be the results of human activity.* First it should be remembered that "anything" could be rewritten as any "thing" or "event" since these are considered interchangeable; however, most past work in the field involves a "thing" conception and terms, and the thing kind of terminology is retained. One notable exception to this traditional view is chronological studies which must conceive data as events for obvious reasons. You cannot date an object before you, since it is still in existence, but, rather, must date some event or events (e.g., the event of manufacture, breakage, deposition). The second aspect of the definition which might require explication is the use of "attribute." Attribute must be understood both as "thingness" and "eventness." Not only is attribute intended to refer to qualities in the ordinary sense of quality, but also to position or location in the three-dimensional world. Human activity is manifest not only in changes of form but also changes or reorganization of locations, and, indeed, is usually a matter of both. One need think only of the importation of raw materials to have numerous examples of artifacts by virtue of location alone. The final aspect of the definition requiring some additional consideration is the "can be assumed to be" phrase. Unless one sees something being modified in form or moved, one must always *assume* the agent of human activity. Since prehistory is most often concerned with the past rather than the present, this becomes an important aspect of artifact and is the reason for the insertion of "assumed" in the definition. It is assumed that a given object or event is a product of human activity if its location or any other of its attributes cannot be accounted for by known natural processes. Thus the identification of artifacts is a problem of comparison with the known products of natural processes. It is important to recognize that individual attributes of objects are not in and of themselves distinctive of human activity until that point in history in which man begins to chemically alter the natural environment. Rather it is *pattern*—on an object, over a series of objects, or through space—that is distinctive. Prior to the advent of constructed materials the only means for shaping stone, for example, were pecking, grinding, and chipping, all of which occur naturally. Much prehistoric literature to the contrary, the removal of a flake is not the basis for assuming that an object is an artifact; however, the pattern of flakes removed from an object or the patterned occurrence of the objects through space may provide such a basis. For example, a chip on a finely-worked Danish Neolithic dagger is not distinctive of human modification. Each flake individually considered could well be the product of natural processes; however, the patterned occurrence of several hundred flakes resulting in the dagger form is distinctive especially in view of the large number of such objects known to occur and the context in which they are found including other objects most easily explained as the products of human activity. Only in those cases in which too little information is available to make appropriate comparisons is there any difficulty in deciding whether or not a given object can be assumed to be the product of human activity.

In this context it might be pointed out that science inevitably sacrifices completeness for accuracy. In viewing the identification of artifacts as a comparative problem, it is important only that everything identified as an artifact be indeed an artifact. Undoubtedly many things will be excluded that should be included, but this is not of pragmatic consequence. One of the normal kinds of progress within a science, and certainly here within prehistory, is the continual expansion of its sufficiency.

It is well to digress at this point to consider the utility, the necessity of theoretical definitions such as that presented for artifact. The several definitions of artifact in the archaeological literature can be viewed as special cases, restrictions for one or another reasons of this theoretical definition, and can be logically derived from it. If two definitions can be derived from the same general proposition, then the relationship between the two can be stated. Special definitions are often adaptations to the contingencies of executing a piece of research. Some definitions specify the scale of the object to be considered an artifact as portable discrete objects. This kind of definition is useful for the recovery and recording of data in the field, for obviously the size and coherence of an object have important bearing on techniques to be used. In this case, other larger or less coherent objects are given other designations such as "features" or "structures." Non-discrete units based upon proximity and association such as "squatting places" and other identifiable loci of specific activities are gaining currency as artifacts. Because of their lack of discreteness, a function of scale, these units must be analytically constructed and thus are terminologically differentiated from the more usual discrete objects. Such "features" and units are artifacts in the same sense as those items given the label "artifact," and they will be treated the same in any system of explanation. The differentiation is simply a recognition of the effect of scale and coherence on recovering and recording data.

Another kind of operating restriction is the division of artifacts into "incidental objects" or "non-cultural debris" or "food remains" and "artifacts." In this case the restriction serves to segregate artifacts into categories requiring different kinds of academic specialists for identification—bones to the zoologist, plants to the botanist, and tools to the prehistorian. Again, all the categories have the same logical properties. The differentiation reflects only the structuring of academic disciplines, not some difference in kind in the data.

Special definitions are likewise employed for particular kinds of problems. For example, an investigator interested in stylistic change might advantageously restrict artifact to intentionally manufactured items. This kind of definition is not at all uncommon in archaeological writing, for style has been an important area of inquiry. An investigator interested in technology may restrict artifact to manufactured items, the by-products of manufacture, and the raw materials. In similar fashion one finds that artifact is frequently restricted to modified forms in studies dealing with early man where the presence of man and his activities is problematic.

All of these special definitions and many more are best treated as part of method, and not matters of theory. All can be derived from the general theoretical definition and related to one another explicitly. If, in the construction of a program of research, the investigator starts with a theoretical definition and adapts it overtly to the problem at hand, the frequently encountered problem of utilizing concepts inappropriately defined for the particular purpose to which they are put is eliminated. Further, a precise statement of the comparability of different studies is possible, and the perspective gained from employing this procedure in developing tactical concepts also aids in recovery procedures. It is unfortunately true that in some parts of the United States many kinds of tools have not been collected in excavation and surface reconnaissance because the investigators were implicitly using a restricted definition of artifact which had been developed in stylistic studies; this has quite effectively rendered the data useless for any other kinds of studies. Most of the argument about what is to be called artifact and what is not is an argument about words, for argument is usually focused on two or more special tactical definitions designed for different purposes. The single most important benefit obtaining with frankly theoretical definitions is that theory—the concepts themselves apart from a particular problem—can be discussed. Indeed, there cannot be theory without such definitions, and with them arguments at cross-purpose can be avoided. Further, laws are impossible achievements until the terms in which they are phrased are theoretical.

Returning to the concept artifact itself, there is one final point that cannot be emphasized too strongly. Defined as it has been here, artifact is the *only* subject matter of prehistory. Prehistorians do not study "culture" or past "societies" or "man's past." Culture and society are anthropological concepts, and man's past, a metaphor. The only tangible phenomenon which can serve as data, with which prehistorians actually work and which is capable of explanation, is that encompassed by artifact. Confusing the means of explanation (culture, society, and so on) with the phenomena that are to be explained (artifacts) only results in further confusion, inconsistency, and untestable conclusions. This, of course, does not mean that one cannot study concepts, or any other words, for that matter; it only means that such study is not prehistory, but rather philosophy or linguistics, depending upon the approach.

Culture

Culture is the most overworked word in the anthropological jargon. It would sometimes seem that every initiate to the anthropological disciplines must invent a definition for it to gain admittance to the profession. In 1952, Clyde Kluckhohn and A. L. Kroeber recorded some published definitions and concluded their treatment with one of their own, summarizing the salient features of previous definitions. Their definition constitutes a generalization, for not all of the features they include occur in any definition. The lack of a generally accepted meaning for the term which prompted the Kluckhohn and Kroeber endeavor appears in retrospect to have been aggravated if not generated by the insistence upon using substantively-bound, special-purpose definitions. The Kluckhohn and Kroeber definition did not rectify the problem. Indeed, this definition probably has less currency than many of the definitions it summarizes. As a generalization it still is restricted to the problems that were covered by the summarized definitions, and is too unwieldy for practical use. The disagreement, inconsistent usage, and outright contradictory content of many of the various definitions has been further complicated by a penchant for including as part of the definition various inferential elements that pertain to why the concept may be useful.

Herein the concept *culture* is to be understood as meaning *shared ideas*—and nothing more. The various special-case definitions may be derived from this by:

    (a) restricting the coverage to some special set or sets of shared ideas, in the fashion that restrictions can be imposed on the theoretical definition of artifact;

    (b) inferring or speculating how the ideas come to be shared (e.g., those stipulating learning);

    (c) inferring why the ideas are shared (e.g., those which view it as an adaptive system, etc.).

These tactical definitions have their place in methods (e.g., a special definition for the problems and views of economic anthropology) and in techniques (e.g., a definition adapted to the particular data being studied). However, they cannot provide an adequate basis for theoretical considerations.

Quite apart from this notion of culture as an explanatory concept, there is the use of culture in the partitive sense in both sociocultural and archaeological literature. In speaking of "a culture" sociocultural anthropologists are denoting a set of people who to a greater or lesser degree share a number of ideas which are not shared by people outside that set. In an archaeological context, "a culture" is even more vague, denoting either a given set of assemblages of artifacts or a set of abstract units such as phases or components, which hold in common a relatively large number of features or "traits." This usage of the term culture, in spite of the label, bears little relation to culture as an explanatory concept and is nowhere employed herein.

There are some important implications, however, of even the simplistic definition of culture used. First, culture is a concept, an idea. It has no objective existence itself and is not subject to study or explanation in any scientific fashion. It is a means of explanation. Further, its referent, shared ideas, does not have any objective existence. Ideas cannot be observed, but are always inferred from behavior, linguistic or otherwise, or products of behavior. A simplistic parallel can perhaps be usefully drawn between culture as an explanatory concept and the concept of gravity in the physical sciences. Gravity is a concept used in the explanation of the motion of bodies. There is no gravity in the phenomenological world; no one has ever seen it, and no amount of generalization will ever lead to gravity. Gravity is a posited concept which permits the prediction of the motion of bodies in fully calculable terms. What is observed is the motion of bodies; what is explained is the motion of bodies, and it is done in terms of the concept gravity. As in the case of culture, the referent for gravity is not observable; that is, forces cannot be seen or measured apart from the motion of bodies. It is in this manner that the concept culture can be and implicitly is employed by prehistory. Arguments as to whether artifacts and/or behavior are "culture" are just as nonsensical as arguments about whether the moon or its motion is gravity.

The character of the concept is imparted both by the stipulation of ideas as a referent and that the ideas must be shared. There is little doubt that explanation of artifacts and behavior can be usefully attempted in terms of the ideas held by the people involved. Perhaps the point of confusion in this respect revolves around *which* ideas of the people are considered. It is obvious that the "ideas" that are solicited from living people under study are not the means of explanation, but are part of what is to be explained. The ideas which serve as the referent for culture are imputed to the people to provide the mechanism for explanation, much in the manner as the force called gravity is imputed to nature for the explanation of motion of bodies. It is unimportant and indeed unknowable if either the forces or the ideas actually obtain in nature. What is of importance is whether or not such concepts permit the development of explanations, for explanations as predictions and means of control are testable. Nothing could be gained from a demonstration that the ideas called culture exist beyond the mind of the anthropologist or prehistorian.

While there has been some criticism of the "sharing" stipulation (see bibliography), this stipulation derives from and is a requirement of a science. "Sharing" implies, or rather is a rewording of, repetition or recurrence through time and/or space of some form. Without repetition explanation is impossible because nothing recurs to be explained. Without repetition both systematics and science are impossible. To conceive of data as unique or "idiosyncratic" is to abandon any attempt at explanation (not infrequently when these terms are used, they are tendered as explanations and employed as a warrant to consider no further the data so labeled). From the outset phenomena are assumed to be unique, and the problem is to categorize them so that they are no longer unique and thus are capable of explanation. Science cannot predict when a given molecule of water will leave the surface of a pan of water and at what temperature, but it can predict, and quite accurately, when the pan of water will boil and how to bring about that condition. In short, the stipulation that culture, as a scientifically useful concept requires a component of sharing, permits the possibility of ahistorical laws for human activity. The concept culture, then, provides the means by which prehistorians explain the products of human activity. Quite obviously, this does not exhaust the possibilities of explaining those products; it does so only for them as artifacts. Physicists, geologists, biologists, theologians, and farmers can also explain the same objects, each with different results. One might insist in the first three cases that the results are trivial because they do not account for the human aspects or because they yield more interesting results for other phenomena, and in the latter two cases one might object that the results are not scientific and do not explain in the sense used here. But certainly none of these are wrong.

In summary, the definition of prehistory tendered earlier in the chapter can be more fully explicated. Prehistory is a kind of study, a science, sharing with other sciences the aim of explanation of phenomena utilizing a theoretical structure. Prehistory is distinguished from the other special sciences in that it employs the concept of culture as the basis for explanation of phenomena conceived of as artifacts.

The relationship obtaining between prehistory and non-sciences is but a special case of those obtaining between science in general and non-sciences which have been already considered in Chapter 1. Thus a treatment here of the relationships between prehistory and humanistic studies in general would be redundant; however, some detailed consideration of the relationship of prehistory as defined here and history and sociocultural anthropology is warranted by the close connection attributed these three fields in some archaeological writing. History, as was asserted earlier, can be distinguished from science, and thus from prehistory, on two fundamental grounds: (1) history does not produce or attempt explanation in the sense of prediction and control; and (2) the organization of history's data is assumed to be chronological. Thus the only "theory" history need employ is a common cultural background of writer and reader. History does not closely articulate with prehistory except in the sound of the name.

That history produces chronicles is not distinctive, for science likewise produces chronicles. However, the chronicles of science must be rendered in terms of classes derived from theory, whereas the historical chronicle consists of chronologically-linked, unique events. The scientific chronicle, much a part of prehistory, is easily confused with historical chronicle, especially in prehistory where the theory employed in constructing a scientific chronicle has been left implicit. To further complicate matters, historical rather than scientific chronicle often appears under the title of prehistory, usually called "culture history." Indeed, most of the things called "culture history" are examples of the non-explanatory "descriptive" approach specifically excluded in defining prehistory as a science. Because the subject matter is usually preliterate man, "culture history" is usually an *inferential* historical chronicle, both the chronology and the events being inferred. In the view taken here, this kind of culture history is properly the practice of history on preliterate data, a kind of ancient history. History and prehistory are not complementary studies in terms of their subject matter. Each is applicable to the results of human activity regardless of the presence or absence of written records, though this feature profoundly affects the techniques of data collection.

Likewise, prehistory is applicable to contemporary results of human activity. The results of this application are less interesting to most people than those produced by history or other humanistic and scientific studies. For this reason, perhaps coupled with a feeling that one is not an "archaeologist" unless one deals with very ancient data, prehistory has seen comparatively little application to contemporary or modern data.

Importantly, history and prehistory have little in common, being quite different kinds of study with discrepant aims and potentials and overlapping fields of application. The general feeling that they are similar stems first from the fact that both make use of the chronicle, though each uses the chronicle differently, and, indeed, the chronicles themselves are different; and, secondly, because most prehistorians are also historians, that is, most people who practice prehistory also at one time or another construct "culture histories." Given their radically different nature, a separation of the two is absolutely necessary if progress is to be made in either.

In the United States, but not universally, prehistory is academically considered part of sociocultural anthropology. While the close connection and in many respects profitable association between the two is not to be denied, it is difficult to conceive of prehistory as a science and also a part of or kind of sociocultural anthropology. Most kinds of sociocultural anthropolgy have little about their nature to suggest the field is a science, though good cases can be made for particular branches (such as ethnoscience) and particular studies being so. Currently, the main part of sociocultural anthropology is more like a flat history of mainly non-western peoples. This does not mean that sociocultural anthropology is incapable of being the science of man, but simply that most of it is not that and is not developing in that direction. There are, however, important connections between prehistory and sociocultural anthropology, far more so than obtain between prehistory and history. The primary point of articulation is in the concept culture, a concept developed by sociocultural anthropology. Sharing such a fundamental concept has naturally resulted in a great many correlative commonalities. In many respects the terminology used to manipulate data is the same. Further, sociocultural anthropology's broad interests in all kinds of human activity have been adopted into prehistory, along with the perspective that comes from familiarity with non-western lifeways. Thus many of the distinctive and essential elements and directions of prehistory are held in common with sociocultural anthropology; however, these are articulated into two different kinds of study. In the case of prehistory the concepts are part of an over-all theoretical system aimed at explanation of human activity, whereas the kind of articulation these same concepts receive in sociocultural anthropology is less systematic, more various, and, at least from the outside, less indicative of a purpose. While one can be appreciative of the important contribution made by sociocultural anthropology to prehistory, there is nonetheless a very stringent limit to the interdependence of the two given their different structures, potentials, and aims.

With the "cultural reconstruction" approach, generally acknowledged to have been given its modern impetus by Walter W. Taylor's *A Study of Archaeology*, there is an attempt to do sociocultural anthropology in the past. Not an insignificant amount of modern endeavor represents a technically more sophisticated and less ambitious version of this general approach. To a limited extent all prehistorians engage in some kind of reconstruction, or, rather, construction; however, as an approach, "cultural reconstruction" has all of the non-explanatory, descriptive limitations of old-time cultural anthropology, complicated by far inferior data. The interests, not the methods, of sociocultural anthropology and its ancient analog "cultural reconstruction" makes a valuable contribution to prehistory. Again the point is not to criticize "cultural reconstruction" as such, but simply to note its exclusion from the realm of science and to differentiate it from prehistory.

In discussing the relationships of prehistory to history and sociocultural anthropology, alternative approaches to the study of artifacts have been indicated. History and sociocultural anthropology are not, of course, the only alternative studies of man's activities. There are many well-developed fields, mainly within what has been called humanistic studies, which attend a more restricted segment of man's activity. Many kinds of study and inquiry have something to offer about the remains of man's past. Prehistory is but one such study, the science specifically directed toward these remains.

The problem to be pursued in the remaining chapters is simply a delineation of how one gets from science in general to a science of artifacts—essentially a substitution of prehistory as defined here for science in the general scheme presented in Part I. The definition of prehistory provides all the necessary elements for making the logical step from science to a science of artifacts. Given the earlier scheme, this is phrased largely in terms of a shift from the arrangement of things, to the cultural arrangement of things. No attempt is made to develop a new means of making this step, but rather the aim is to make explicit the implicit manifestation of this step in the literature of prehistory.

6

CLASSIFICATION
IN PREHISTORY

*I*n order to draw directly upon
the propositions explicated in Part I, it is necessary to be able
to treat prehistory as a special case of science, as a distinctive
restriction of the general field. The definition of prehistory provided in Chapter 5 permits this kind of derivation by stipulating
the kind of restrictions required to convert science, an abstract
notion, into the science of artifacts, one of the several special
sciences. Drawing upon that definition, prehistory can be viewed
as science restricted to the explanation of artifacts in cultural
terms. Systematics, the means of formulating units and the subject of discussion in Part I, is held in common by all the special
sciences; however, the specific form of the units employed and
the kinds of choices and decisions made in their formulation are
distinctive for each science. It is the particular kinds of arrangement of phenomena, governed in form by the theory of the special science, that provide the basic material for the science and
its operations. *Phenomena categorized for use by a specific science are* customarily called *data,* and the term data will hereafter be restricted to such categorized phenomena. Phenomena
will be retained for things and events without such categorization. In the widest sense, the data of prehistory are artifacts.
Since the means of segregating artifacts from other phenomena
was necessarily discussed in defining prehistory, the problem to

be considered now is how these data are structured for explanation by prehistory. Given our restriction to formal theory, the logical transition from science in general to a science of artifacts is a matter of the derivation of cultural arrangement from arrangement in general.

Systematics is necessarily part of any scientific endeavor, though it is rarely the focus of that endeavor. The means by which the units used have come into being and how they are identified in the phenomenological world are usually implicit, the investigator having learned implicitly what has been traditionally employed. Evaluation of those units is even less common than an explicit presentation of them. Prehistory represents no exception. This deficiency is of far more import for prehistory than other physical sciences because, as has been indicated, the subject matter of prehistory cannot be viewed as something external to the investigator. The investigator is part of it, and so is his work. The temptation for him to use his own cultural background as theory for creating and manipulating units, rather than treating this background as subject matter, is great and deleterious. Explicit systematics, however, enables the prehistorian to separate his cultural background analytically from the theory employed in his investigations, as well as to make poorly expressed or unexpressed theory explicit.

There are, of course, important exceptions to this malady of implicitness in the literature of prehistory, works which thoughtfully consider systematics in relation to both phenomena and problems. For a variety of reasons they have not, however, been overtly and systematically employed either by the substantively-oriented majority of prehistorians or the increasingly large body of statistically-oriented tacticians. In spite of the paucity of overt use, Irving Rouse's *Prehistory of Haiti: A Study in Method* and J. O. Brew's *The Archaeology of Alkalai Ridge*, along with two articles, Alex D. Krieger's "The Typological Concept" and Albert C. Spaulding's "Statistical Techniques for the Discovery of Artifact Types," form the implicit basis of almost all of the archaeological literature that might be called prehistory. It is difficult to assess whether these works have actually, in an historical sense, been the derivation of the units employed in the literature, or whether they are simply overt expressions of a pre-existing but implicit approach by prehistorians. Regardless of this point, the traditionally employed arrangements of prehistory are understandable in terms of the notions advanced in these studies. Perhaps the most remarkable element is that none of them has led to any measurable increase in the explicitness of systematics in the discipline. Primary among suspect causes of this condition is the fact that none presents a wholistic scheme completely free of substantive connections. Further, each, to a greater or lesser extent, is unnecessarily infused with inferential properties. It has been these inferential aspects that have received elaboration by the authors and subsequent students, and these same aspects which have suffered severe and justified criticism and rejection. The article by Spaulding is concerned both with grouping of the first kind discussed in Chapter 4 and with classification, while the studies of Brew, Krieger, and Rouse are primarily within the field of classification. By necessity, these latter three studies and their subsequent elaboration form the basis of the examination of classification in prehistory undertaken here, as the Spaulding reference forms the basis for grouping in the succeeding chapter.

Within those arrangements distinctive of prehistory, classification plays the crucial role in the transition from science in general to a science of artifacts, for, as has been shown, classification is the only means of creating the intensionally defined units necessary for science. These units, as in other kinds of science, become the data in that they subsume all the relevant attributes of the phenomena for the particular kind of inquiry represented by prehistory. Further, they provide the terms by which the data can be discussed and manipulated. It is useful, then, to lay out the specifications that all classification must meet for prehistory in general terms before treating more specific forms.

Recalling the earlier considerations of classification, it is necessary that the field for the classifications be defined, along with the problem to which the classification is directed and the attributes to be used in creating classes. At this most general level of concern, the field is that encompassed by the concept artifact, objects which owe some of their attributes to human activity. The problem similarly is to provide categories for these data that are cultural, for the ultimate purpose of explaining the products of human behavior and with them the behavior that created them in terms of ideas held in common by the makers and users. It should be re-emphasized here that location in the three-dimensional world is an attribute of an object as much as its color. Quite obviously some additional assumptive elements are required beyond those necessary for the construction of classification in general in order to derive classifications which meet these special stipulations. Indeed, the soundness of the formal basis of prehistory, and thus of prehistory as a branch of scientific inquiry, can be assessed from (and is a function of) the number of additional assumptions that must be made.

The additional assumptions are introduced in specifiying the general characteristics that features used to create classes must display. The general field from which definitive attributes may be drawn is implicit in the notion of artifact. *Only* those attributes which can be assumed to be the result of human activity are useful. The identification of such attributes is a product of comparative study similar in all respects save scale to the identification of artifacts themselves. Stipulation of the appropriate field of attributes insures that objects identified as products of human activity will be further structured as products of human activity. For example it is quite possible to use a set of artificial attributes, either intuitively or overtly, to identify an object as an artifact, but then to further categorize the object as to kind in terms of natural attributes present on the object only incidental to its nature as an artifact. A clam shell as an element of a coastal shell midden can be readily identified as an artifact and this is generally done, though not necessarily under the term "artifact." It is quite possible, however, that the total sample of clam shells might be categorized in terms of color, resulting in brown mussels, white mussels, and brown-and-white mussels. If the differences in color are due to differential preservation of the outer horny layer, the use of color as a dimension of features is clearly erroneous in a cultural classification. These kinds of errors are avoided by an explicit statement of the general requirements that must be met by an attribute for the purposes of prehistory.

It must be emphasized that the suitability of any set of attributes must be determined for each particular case as a product of a comparative study. No absolute list of attributes can be drawn up and labeled "relevant" or "cultural." The attributes which can be shown to be relevant will differ from case to case. The material from which artifacts are made provides an excellent case in point. Within the realm of stone artifacts, the chemical composition is unmodified—only the shape is changed. Nonetheless, the chemical composition is frequently cultural, being the product of selection manifest as artificial locations. Often the relevance of chemical composition goes no further than its effects on whether the material will chip or crumble upon impact, a simple two-feature distinction. Only with detailed comparisons with the environment, however, will one be able to ascertain whether this simple set of features is adequate or whether a more complicated set involving color, texture, hardness, etc., are involved in the selection. The occurrence of *only* sandstone rocks as heating elements in earth ovens, when both limestone and sandstone are available, indicates that one was given preference and that the set of features used in creating a classification for this material must differentiate the two. Anyone familiar with the characteristics of these materials when heated will readily appreciate the reasons behind such a preference. Again, the point is that no absolute set of features can be set forth as universally relevant. Much in the same manner that a linguist must convert his phonetic record of speech into a phonemic record which is cultural, the prehistorian must demonstrate by comparison the relevance of the features to be used.

The use of the term "cultural" to mean relevant for explanation in terms of the concept culture is premature at this juncture for the definition of culture not only stipulates the element of human involvement (ideas) but restricts this general field to that set of ideas which can be assumed to be *shared*. This is a most crucial point, for it is here that the articulation of phenomena with concepts is made. This connection necessarily must be made by means of assumption. There are no articulations between the abstract and the real which are observable or demonstrable. Clearly, the assumption made is the formal foundation for all of prehistory, constituting the means by which science becomes the science of artifacts and serving to differentiate prehistory from other sciences. While there is no overt consideration of this point in the archaeological literature, it is implicitly considered in many works and the nature of the assumption is quite clear. *Prehistory assumes that attributes which are the products of human activity and which recur over a series of artifacts* (termed features) *can be treated as manifestations of ideas held in common by makers and users of those artifacts.* Thus the link is made between the phenomenological and the ideational. In spite of its simplistic appearance this assumption has several ramifications which require exploration. Because it is the basis of all prehistory, the reasonableness of the assumption must be questioned.

The importance of restricting the possible set of attributes to those which are demonstrably products of human behavior is evident. If the attributes considered are *only* those which are the products of human endeavor, it follows that any explanation of those attributes is necessarily done in human rather than natural terms. If their distinctiveness lies in their humanness, then so does their explanation. Further, given our assumptions about the uniqueness of the phenomenological world, recurrence or sharing necessitates an ideational element in the explanation. Some kind of classification is required as the vehicle of explanation. If several objects hold features in common, and those features are of human origin, there is but a single plausible account: Intentionally or unintentionally, consciously or unconsciously, the objects were made to look alike by people who can be treated as possessing similar ideas about them and who have the same categories of features and ways of articulating the features into whole artifacts. In short, the objects can be treated as expressions of the same mental template. Now obviously this connection can be challenged in any given case by special explanations utilizing natural processes and chance; however, given the large series of cases represented by artifacts, infinite for all practical purposes, such challenges are trivial. No other *single* account is capable of subsuming *all* of the cases at hand. Nonetheless, given both the language available for stating the assumption and the discussions presented in archaeological literature, important potentiality for misunderstanding the assumption exists. Three aspects need to be made abundantly clear in order to avoid any serious misunderstanding: (1) the locus of shared ideas; (2) the means by which they are shared; and (3) the scale at which they are shared. Each of these is treated briefly below.

1. LOCUS. While it is common to impute, at least for literary convenience, the sharing of ideas to the makers and users of artifacts, clearly this cannot profitably be demonstrated or held to be true. Ideas are not observable—only behavior and its results are. There is no way to know what, if anything, goes

on inside a living person's head, let alone a dead one's. The "sharing" element lies in the process of converting unique attributes into features which can recur, a process done by the prehistorian as the intuitive first step in analysis. What is important is that the recurrence of features over a series of objects *can be treated as if* there were such a force. As long as the units are systematically tested against phenomena, there is no point to querying whether or not the makers used the same categories as the investigator did, for the testing insures that the same end product is reached regardless of the route taken to get there. It is immaterial, for example, whether in learning to identify plants in some exotic language you use the same criteria as do the native speakers, so long as whatever criteria you do use produce the same assignments. There is no way to demonstrate that your criteria are the same as those of a native speaker or that the natives even share among themselves a single set of criteria. One thing that this discussion does indicate quite clearly is that "culture" is implicitly used by prehistorians, at least in the initial stages of classification, as other explanatory concepts are used in the physical sciences.

Since there has been some attempt to link the classifications of prehistory with the "folk classifications" of subject peoples (principally in cultural reconstruction approaches), some consideration of this specific aspect seems warranted. Above it was argued that this kind of linkage is unnecessary. Further, because it can never be a matter of demonstration, to make this a criterion of "good" classification is to base prehistory upon an unprovable and untenable proposition. The only utility in asserting that the locus of sharing is in the classification instead of the subject matter of the prehistorian is to eliminate this undemonstrable proposition; otherwise, and for practical purposes, the question of locus of sharing is trivial. It is further useful to indicate not only that linking "cultural classification" to "folk classification" is unnecessary and unparsimonious, but also that it is detrimental to the purposes of prehistory. Folk classifications, when such are obtainable, constitute subject matter like any other artifact or behavior instead of units of analysis and synthesis. To use folk units as the units of study is not terribly unlike a taxonomist asking a frog to what species he belongs. If an attempt is to be made to understand frogs to a greater degree and in a different manner than frogs understand themselves, the

frog's answer is going to be treated as an instance of highly un-
usual behavior and not as a scientific unit. The source of this
latent tendency in prehistory to regard as an ideal a congruence
between cultural classification and folk classification is undoubtedly sociocultural anthropology, where many "analytic"
units such as the named social units are elicited from the people
themselves.

The potential problems that can arise from such an equation become obvious if the temporal dimension is considered.
How can one study change through time—say of projectile
points—using a folk classification for projectile points current
in A.D. 1, when that classification can hardly take into account
projectile points made in the following 2000 years? Further, the
classification, as a cultural phenomenon, changes through time
as well as the phenomena it serves to order. The definitions of
the classes will gradually change in meaning, introducing the
very ambiguity that analytic classifications are intended to eliminate. The flat temporal perspective of sociocultural anthropology
admits this kind of error more readily than does the context of
prehistory. When time is meaningfully introduced, the equation
between "folk classification" and "good cultural classification"
is negated. The nature of folk classifications as grouping becomes
apparent. As groups, such devices are restricted to a finite realm
of time and space and to the particular view of that realm taken
by the persons using it. The common sense categories of English
are exactly the same. Attempts to categorize data with such
"rubber yardsticks" can hardly be expected to yield meaningful
units in any scientific sense. The rejection of grouping in general
and folk classification in particular as a means of creating units
for prehistory is not intended to exclude the latter from study.
As a means of study they are useless, even deceptive; as a subject of study they may offer a great deal.

2. MEANS. The assumption posited as the basis of cultural classification does not stipulate the means by which ideas
come to be shared. Indeed, whether or not ideas are actually
shared is a trival point. The sharing or recurrence of features
is a function of classification and thus is purely formal. Many of
the considerations in the literature are crippled by *inferring* the
means of sharing, thus forcing the foundation of cultural classification to rely upon inference. These inferences are usually

focused on distinguishing functional resemblances (that is, those features which are common to sets of artifacts because they were used for the same thing), from stylistic or historical resemblances (that is, features held in common as the result of historical connection either contemporaneously by diffusion or traditionally by persistence of style). Both of these assessments are obviously inferred from the observation of a feature's distribution over a series of objects, sharing in a purely formal sense. No doubt there are components of both functional and historical resemblance in the configuration of almost any object, so that, further, the inference is one of degree. Sharing as used in these discussions is formal, implying neither historical nor functional means of sharing. The means of sharing has to be inferred from the number, pattern, and distribution of the shared features; it is a problem to which some attention has been directed but is not part of formal prehistoric theory.

3. SCALE. The third feature of the assumption is that no scale is specified for recurrence or sharing. The terminology used perhaps implies recurrence at the level of attributes of discrete objects; however, this is but the most commonly employed scale in prehistory. The units which share features need only be readily bounded in the phenomenological world. Thus the units may be communities, with the features, as house types; the units, houses, with the features as constructional elements of houses; the units, house floors, with the features as elements of house floors; the units, hearths, with the features as parts of hearths; the units, hearth lips, with the features as elements of hearth lips, etc. Only a relationship of scale between units (which must be bounded phenomena) and features (which must be classes of attributes of those phenomena) is stipulated. While the practical problems of discovery, recovery, and recording certainly do vary with the scale, the logical properties do not and thus have no role in theory.

This consideration of scale in relation to sharing brings into focus the contrived nature of the cultural/idiosyncratic contrast briefly noted in the preceding chapter. First, sharing is purely formal and inheres in the classification, and is not an intrinsic quality of phenomena. Adding to this the lack of intrinsic scale, one can easily appreciate that the question of whether or not two objects share features is a direct function of the definition of the features and the scale at which they are conceived. Two objects share or do not share features dependent only upon the discriminations made by the investigator. For example, two houses may be different in structure, one being built on piles, the other being built on the ground; one being small, the other large, etc. They may be regarded as different on these bases, and, if the pile-house is the only example of such a structure in a sample consisting otherwise of ground-level houses, it might be called idiosyncratic. It is idiosyncratic only in terms of the features used in the judgment. A different set of features, such as construction materials, function, etc., can be used to group the two structures together as the same thing. The two houses may be different as houses, but identical as parts of houses; that is, they differ at the scale of "house," but are the same at the scale of "part of house." Each house is made up of different arrangements of the identical features or parts. Any two objects which do not share features may be made to share features by reducing the scale of the comparison to parts of the objects. To call one object idiosyncratic because at a different scale, usually unspecified, a particular feature or set of features is not held in common with some other specified set of objects, is a failure to grasp the problem or the potentiality of classification. The relationship obtaining between two objects can be precisely specified by a statement of the nature and number of features held in common *at a given scale*. That at a given scale a specific set of features is not shared is perfectly evident, and the "idiosyncratic" object clearly differentiated, but not as something apart from a cultural system and unamenable to further inquiry utilizing cultural theory. There is a strong tendency, not only with the idiosyncratic/cultural dichotomy, to "freeze" scales and treat scale not as customary, but as absolute. The reasons for this are simple. The terminology is a product of such customary investigations, and each term is linked to either features or units at a given scale. Theoretical terms are lacking. While the terms "unit" and "feature" may lack appeal as "jargon," they do permit one to discuss sharing and the units shared, as well as the vehicles of recurrence. The basic assumption does not and need not specify any scale. This needs to be specified for particular techniques and methods, but except as a concept scale does not enter into theory.

In summary, then, the assumption made by prehistory

equates recurrent features of human origin with shared ideas of the makers and users of artifacts which display such features. This assumption is implicit in the literature of prehistory as a general proposition, though corollaries derived from it as statements at specific levels and for specific purposes are sometimes explicit. The assumption utilizes shared ideas as an explanatory device—it is not necessary or even desirable to hold that shared ideas, culture, are actual constituents of the phenomenological world, any more than insisting that gravity is a force in the physical universe instead of a concept for the explanation of the motion of bodies. While it has been necessary to consider the issues of the locus, means, and scale of sharing, an explicit statement of the underlying assumption as a general proposition avoids the errors made in these areas. Sharing is a formal device and a function of classification. Some kind of sharing or recurrence is necessary for any classification or arrangement and the assumption simply specifies the rules for insuring that the resultant units are useful for cultural theory. Recognition that the means of sharing, functional convergence or historical contact, is an inference based upon, not a part of, observable formal recurrence patterns, eliminates the second area of concern. Finally, a recognition that what is cultural, that is, what is shared, is a function of the scale of comparison as well as the features and units themselves, and thus relative, eliminates arguments based upon absolute statements of what is cultural, such as involved in the idiosyncratic/cultural dichotomy. The assumption posited as the formal basis of prehistory functions to derive cultural classification from classification in general; it provides the means for insuring that the units created are useful for manipulations in terms of the concept culture. It is the link between the scientific systems of prehistory and the phenomenological realm. Utilizing this general background to cultural classification, it is possible to see how cultural classification is actually realized in the discipline, first in terms of the kinds of classification employed, and then in terms of the scales at which it is customarily practiced.

Kinds of Classification

Save in those studies which have arrangement as a goal for its own sake, it is obvious that a kind or kinds of classification are widely employed in prehistory. Both explicit statements outlining procedures and emphasizing the importance of units over the objects grouped in them and the characteristics of archaeological units generally (e.g., their ability to recur through time and space) make this clear. Differentiation of groups of artifacts from classes for artifacts is in evidence in the literature of the nineteenth century and has had overt expression in American prehistory at least since 1939, when Rouse clearly makes this distinction in theoretical terms in *Prehistory of Haiti*.

Identification of the kind or kinds of classification employed in the literature is not an easy matter. Far more frequently than not, classification as a process is implicit, the reader being privy only to the results. Further, it would seem, the process has not been explicit in the minds of many writers, for there are frequent errors of consistency and form. By far the most common and distressing error from a reader's point of view is a failure to differentiate classes from *denotata* of classes. Definitions, as necessary and sufficient conditions for membership in a class, are not presented separately from descriptions of a particular set of *denotata*. This combines into a single undifferentiated mass the features which objects must display to belong to a given unit and the features which the objects assigned to the unit happen to display in various frequencies. The results of using a classification to identify objects is presented, but the classification used is not. For example, the often-encountered "type description" usually consists of a list of dimensions (e.g., in the case of pottery, temper, paste, surface treatment, decoration, etc.) which have been filled in with specific features (shell or limestone temper, regular paste, plain surface, incised decoration, etc.) for each "type." There is no way to differentiate those features and dimensions which an object assigned to a given type must display from those features and dimensions which an object may display. The use of the term "or" as in "shell or limestone temper" is a certain clue to the identification of that dimension as non-definitive. More difficulties are presented with the use of "usually" or "commonly" in deciding whether the features in question are distinctive of a type or not. Comparison with other "type descriptions" in the same set may further enable one to identify dimensions of features which are definitive and descriptive respectively. The lack of consistency resulting from an intuitive approach to classification leads to noncomparability of features used in "type descriptions" such that the dimension of decoration, for example, may be rendered as "incised decoration" in one instance but as "geometric decoration" in another, completely frustrating an attempt to reconstruct the classicfication that has been used. The "type descriptions" are in reality unstructured description of groups of artifacts which have already been identified with classes in a classification which has not been presented. Much of the non-replicability associated with the use classification and classes in prehistory stems directly from this problem—no classification has been presented even though one has obviously been employed. Unless one is willing to practice ethnoscience on the literature of prehistory to reconstruct classifications from unstructured descriptions of sets of denotata, the utilization of such "type descriptions" becomes an esoteric and mystical art. This condition is hardly desirable when the onbly justfiable purpose to classification is the creation of units with explicit, unambiguous meaning.

The obvious, though frequently inconsistent and poorly explicated, use of dimensions, and a lack of overt weighting of one dimension over others, are convincing evidence that paradigmatic classification lies behind most of the units employed in prehistory. Almost all of the kinds of classification labeled "typology" (not all things labeled typology are classification) in prehistory are paradigmatic classification. Regardless of whether the aim is actually achieved or not, a casual survey of any amount of archaeological literature shows that writers intend classes to be identifiable by reference to a set of distinctive features, thus indicating that classification, and not grouping, is being used, and that the features are unordered in terms of identification, thus demonstrating that the classification is paradigmatic. Additionally, classification rather than grouping is indicated by the fact that most archaeological units have distributions rather than locations. Because of poor explication and inconsistencies, this intention is often realizable only to an author and not his reader. The best explicit statements, both in principle and in example, are presented by A. C. Spaulding in "Statistical Techniques for the Discovery of Artifact Types" and by James Sackett's 1966 elaboration of this work in "Quantitative Analysis of Upper Paleolithic Stone Tools." This is somewhat paradoxical in view of the fact that in neither case is paradigmatic classification the focus of attention. This kind of classification is so frequent that it is more feasible to examine those instances where paradigmatic classification as the underlying classificatory device is not assumed to be a sufficient account. This means, given our two-fold division of classification, an examination of taxonomy.

The term "taxonomy" is frequently used to cover a variety of things: a synonym for classification including paradigmatic classification, to distinguish it from analysis; a synonym for what is herein labeled "numerical taxonomy," presumably because of this device's hierarchic structuring; and a label for taxonomic classification. Insofar as it is recognizable, the first kind of usage is unimportant; the second is considered under grouping devices in the succeeding chapter. The only real concern here, then, is the use of taxonomy as taxonomic classification. While the term has been borrowed from the biological sciences, most prehistorians readily agree that prehistory does not have *a* taxonomy comparable to *the* Linnean Hierarchy, nor does it approach its subject matter in the same fashion. The oft-cited reason is that cultural processes are not unidirectional and thus are more complicated than those of genetics and inheritance. While one may allow this as true, it does not have any bearing upon the use of taxonomy—indeed, one might argue that taxonomy ought to be used for these very reasons. The use of taxonomic classification is and has been on the wane in prehistory for some time, largely as a result of Krieger's convincing arguments in "The Typological Concept" against the weighting of features. His arguments are phrased in terms of the practical difficulties encountered in making the required decisions, difficulties that are inherent in the unparsimonious form of taxonomy. In recent literature, taxonomy has played no important role. Some "type descriptions" which are inconsistent in the application of dimensions (the "incised decoration"/"geometric decoration" instance) might be viewed as taxonomies in which only the lowest level taxons are explicit; however, this is probably more a function of an analysis of the sets of "type description" than it is of the classification used by the original writer.

Otherwise, only simplistic sorts of taxonomy are used. The most common form is a kind of additional process in which one begins with an index or set of classes created by the intersection of two dimensions of features. Subsequently, one or more dimensions of features, either singly or in sets, are added, effec-

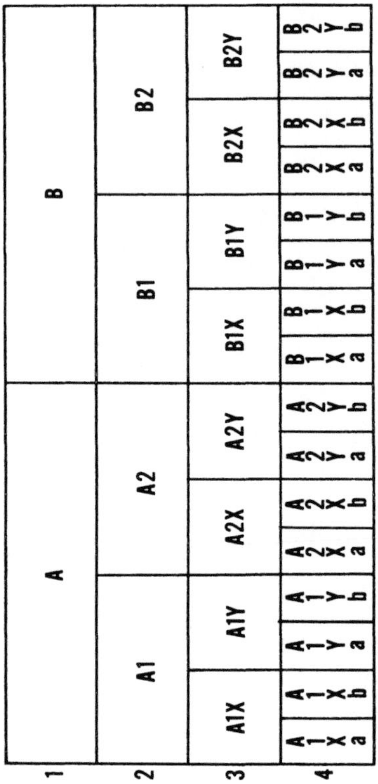

Figure 13. A special-case taxonomy combining the dimensional aspect of paradigmatic classification.

tively "sub-dividing" the initial set of classes. In practice, of course, one could start with the most complicated level and successively remove sets of dimensions—essentially the reverse of the first situation. In prehistory the "ware" and "type" classifications for pottery, frequently used but infrequently explicated, and the "type–variety" classificatory schemes, are of this sort. Figure 13 illustrates the basic design of such a program in which the highest-level classes constitute an index, the second level of classes is created by adding a second dimension of features, the third level is created by the addition of still another dimension, and the fourth level of classes is created by the addition of a final dimension of features. In order to keep the illustration simple, each dimension is divided into two features, but obviously this is not necessary and certainly not usual. Further, as just indicated, this same figure could be described starting from the lowest level and talking about the others as successive subtractions of dimensions. In either approach to description, Class A1 is a kind of A, as A1X is a kind of both A1 and A, and so on. Upon close examination, not only do the classes included under the same superclass at the same level constitute a paradigm, but each entire level is a paradigm. If, for example, one is concerned with only the lowest-level classes, the entire classification *can be treated* as a paradigm. Clearly, then, this sort of taxonomy is a special case within the general field of taxonomy. If any given level in such a device is of concern to the exclusion of others, it is not necessary to treat the various dimensions as ordered or the classes as taxonomic. Class A1Xa can be derived regardless of whether the X–Y dimension is employed before or after the A–B dimension. While ordered, the order is not necessary to derive the classes at any given level.

A legitimate question then arises as to why this kind of device should be regarded as taxonomy rather than paradigmatic classification. The answer is that while any level of classes can be regarded as a paradigm, the entire structure does not present all possible permutations of the features and dimensions, and thus the occurrence of specific classes is conditioned by the ordered addition or subtraction of dimensions. For example, in Figure 13 the occurrence of Classes A1–B2 is a function of applying the 1–2 dimension before the X–Y dimension or a–b dimension. Had the a–b dimension been the second employed in this example the second level of classes would be defined as Aa, Ba, and Bb, and Classes A1–B2 would not occur in the new classification as Aa–Bb do not occur in Figure 13. Clearly, dimensions are ranked in terms of importance, but the features within the dimensions are equally relevant for all previous distinctions. This special-case taxonomy, differentiated from other taxonomies in the consistent and exhaustive application of features through a given level, thus *eliminating the assumptions of position required in other taxonomies,* is potentially a powerful means of unit creation if rigorously executed. Potentially, however, is the key word. While the number of assumptions or weighting required is reduced by the consistent and exhaustive application of each dimension of features, assumptions of importance are still required to order the application of dimensions relative to each other. Unfortunately, the rationale for such decisions is inferential as is the case with the "type–variety" classificatory scheme, and thus the definition of units used to make the inferences depends upon the inferences, a kind of circularity characteristic of taxonomy. It is necessary to be able to answer why dimension A–B is applied first, 1–2 second, and so forth, in terms of observed fact, in order that the taxonomy be sufficiently parsimonious as to be useful for some specified purpose.

The "ware–type" and similar two- or three-level taxonomies, when constructed for a specific rather than descriptive purpose and when the relevance of the features employed in definition is demonstrable (requirements of all kinds of classification), meet this test. Utilizing the lower two levels of Figure 13 as a model, types in the "ware-type" scheme are equated with Level 4 and wares with Level 3. The larger the number of definitive features required of each class, the smaller the distribution the *denotata* of the class will be. Thus, for many kinds of problems, the Level 4 classes are optimal; however, the utility of any set of classes must be weighed against the data being manipulated. As is usually the case with wares and types, the wares represent the fabric of the ceramic (Features A–B, 1–2, and X–Y represent hardness, texture, and temper), and the types include the additional dimension of surface treatment. In practical terms, the fabric of the ceramic is almost invariably recovered with any sherd, whereas surface treatment may often be missing through the agency of erosion. An investigator using a "ware-type" scheme of this sort then has two alternatives available to him, wares *or* types. If his data are well preserved, he will probably employ types. If his material is poorly preserved he may choose wares for this will effectively increase the size of his sample and the reliability of its distribution. In short, the taxonomy provides *alternative* sets of classes, one which makes a maximum number of discriminations but requires optimal circumstances, and another which makes fewer discriminations under less than optimal circumstances. This special-case taxonomy functions, then, to adapt theoretical devices to actual bodies of data, and is really a part of technique rather than theory. The linkages between levels are observational: Surface finishes occur on pastes. The order is likewise observational: Surface treatments are destroyed before the paste disintegrates. Many similar examples of this kind of taxonomy functioning in this specific role may be found in the archaeological literature. There is no reason why more complicated structures cannot be employed for more complicated technical problems.

Further, this special-case taxonomy can be employed in adapting classificatory units to the requirements of particular methods utilizing this same feature of variable numbers of co-ordinate features employed in the several levels. The fewer criteria required for membership, the larger the number of objects which will fulfill the conditions of membership. Thus, using the type-variety method as an example, the level of wares will have greater utility in comparisons through larger amounts of time and space than will types or varieties, and generally are used for such purposes. Varieties, on the other hand, with a larger number of necessary features will be restricted to smaller amounts of time and space and thus are employed in inter-site comparison.

The important point, however, is this: The utility of this special-case taxonomy comes from its characteristics of linked paradigmatic classifications rather than its taxonomic features. Indeed, in the case of the type–variety system, the linkage is observational, and it is this feature, minimizing the taxonomic element, which makes it useful. True taxonomies play no role in prehistoric theory, for to make them parsimonious they must be articulated with the phenomenological realm, and the articulations must be tested as hypotheses. For this same reason taxonomic classifications do function in the realm of technique which attends the articulation of classification and phenomena. The use of paradigmatic classifications linked together with a taxonomic structure is an excellent solution, so long as the taxonomic linkages are not inferential. Those few taxonomic classifications which are based upon inferential notions of "relatedness" or which base the ordering of levels upon inferences about the social groups making the ceramics require the demonstration of such inferences, and such demonstration is presumably the purpose for which the classification is created.

Scale and Classification

Up to this point and in the archaeological literature generally, the terms "level" and "scale" have been used almost interchangeably. It is necessary, however, to differentiate two notions of ranking or inclusiveness treated under the labels of level and scale to further specify the nature of classification as employed in prehistory and the particular kinds of classes that are customary. Implied in the use of both level and scale is a relative degree of inclusiveness or rank. Hereafter, level will be employed to denote inclusiveness in theoretical units, essentially the number of definitive features in a *significatum*. *A level is a set of units (classes) which display the same or comparable degree of inclusiveness or rank.* All the classes in paradigms are of the same level since all are mutually exclusive alternatives with equivalent definitive features in each *significatum*. On the other hand, taxonomies and the special-case taxonomy illustrated in Figure 13 consist of several levels. In taxonomies the level is determined by the number of oppositions and thus the number of definitive features in the definition of a taxon. Being ideational in nature, specific values cannot be assigned to levels apart from other levels. It is thus useful to employ the notion only when two or more sets of units or concepts of differing degree of inclusiveness are being employed, as in taxonomy. Further, the notion of level is applicable only when the various sets of classes constitute alternative classifications for the same phenomena. Types, wares, and varieties are best discussed as classes at different levels, since they differ in the size of the classes produced (inclusiveness from large to small) and since they are alternative classifications for potsherds or other discrete objects.

Scale, on the other hand, will be used to designate inclusiveness or ranking in the phenomenological realm, and thus *is defined as a set of objects (group) which display the same degree of inclusiveness or rank.* Scale is the stipulation of the size of phenomena being considered. One can construct classes for aggregates of objects, discrete objects, or parts of such objects. Although it is not so done, one could construct wares, types, and varieties of all of the various scales just listed. Figure 14 illustrates the scale and level relationships among a series of units to be discussed in later sections. Here the vertical axis indicates scale and thus the relationship between mode, type, and phase is one of scale (they are classes for different scales of phenomena), whereas the horizontal axis represents level and thus the relationship between variety, type, and ware is one of level (they are alternative increasingly inclusive classes of the same phenomena). Level alone is sufficient to discuss classification as a process both in a general sense and within the confines of prehistory. Scale is necessary to specify particular classifications and kinds of units employed in prehistory, and, because it is phenomenological, scale can be specified in absolute terms.

In these terms the concept artifact designates the synthetic level of cultural phenomena. As defined, artifacts have no scale.

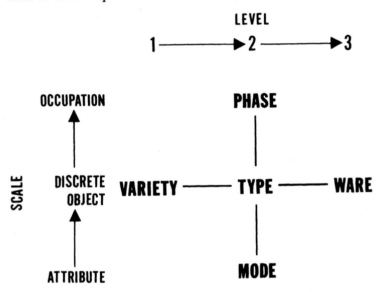

Figure 14. Relationship of level and scale.

Scale is specified by the investigator; it is not inherent in phenomena. A choice, which ultimately must be justifiable, is made. In prehistory the scale of phenomena considered is traditional or customary. This is a most important point. The consideration that follows treats the *customary* scales considered in prehistory. It does not mean, nor should it be construed to mean, that the three scales identified are the only ones possible or that they are the most profitable. There are, however, persuasive arguments in favor of at last a portion of the scales traditionally used.

Implicit in the wording of the preceding discussion and in much of the archaeological literature is a scale best designated as that of portable discrete object, identifiable in that when moved, its component parts remain in the same spatial relationship to each other. A hammer, a coffee cup, and a dog are all examples of this scale. The strongest arguments in favor of using this manipulatory criterion as the starting point in reckoning scales is the ease with which it is identified and the untested but rather reasonable assumption that manipulation of objects is a relevant factor to all living things. Some problems do inhere in the fact that discreteness, like any other quality, changes through time. To take a pronounced example, a dog can be observed to become several discrete objects after death occurs and chemical decomposition begins. After decay eliminates all of the soft parts, discreteness again becomes fairly stable since decomposition affects the bones more slowly. Discreteness and chemical decay are obvious concerns of any investigations directed toward the past. Chemical decay is but one readily observable and familiar form of changing discreteness. As a result of this difficulty, a choice has to be made in distinguishing discrete objects in prehistory: Are discrete objects those objects currently meeting the criterion of manipulatory discreteness, or should discrete objects be considered only those objects or sets of objects which met this criterion at the time at which they were made or used? Fortunately, the former position seems to have been almost universally settled upon by prehistorians (excepting some minor areas of fuzziness), though not without some nagging concern about the changing nature of discreteness. The answer can be weighed as fortunate, for this position permits scale to be determined observationally rather than inferentially. Former discrete units are subjects for inference, but one which is made upon the observed discrete objects remaining to us.

One set of superficial exceptions might be noted, usually presented in the literature under the term "features" or "structures." These units, while differentiated by their label from portable discrete objects, are not treated differently in any essential fashion. The only point of difference lies in the pragmatic realm of recovery; the discrete objects called "features" are not portable but are usually represented by characteristics of soil which cannot be moved without destroying the discreteness of the object or which are simply too big to be conveniently moved. Houses, pits, and fire hearths are objects which usually fall into this category. While differently labeled and frequently described in separate sections of reports, they are usually treated as objects equivalent in scale to potsherds and projectile points.

The discrete object is the basis for reckoning the other two commonly-employed scales. Even the casual student of archaeological literature is aware of a scale larger than that of discrete object, if only because discrete objects are often treated as component pieces of larger things. One looks, however, in vain for an explicit statement of what the scale is or how it may be identified. There is, for example, no explicit statement of what "phases" are classes of, though their nature as classes is perfectly obvious. More frequently than not, tautology characterizes statements relating such classes to the phenomena which they purport to order, running something like "phases are classes of components" and then "components are manifestations of phases." In this case one must have the classes to recognize the phenomena, and one must have the phenomena to construct the classes. While there are numerous classifications in evidence at this scale, there is nothing in the literature to suggest that the inventors of such classifications know in any precise way what they are classifications for. The scale of phenomena is simply not identified. This is, without any doubt, the most serious deficiency in the formal theory of prehistory today.

There are a number of contributing factors. Our own perception of phenomena dictates that any scale larger than discrete object will be seen as a group of objects, some kind of aggregate. In a real sense the phenomena are constructed, and thus the possibility of different aggregates' being constructed by different people exists to a degree not possible at the scale of discrete object. Further, and unlike the discrete object, the dimensions of time and space are apparent in aggregates. What is lacking in prehistory is a statement of how such aggregates are to be constructed. Lacking a common perception and lacking special rules to overcome it, prehistorians have created, largely by accident, a Pandora's box of phenomena, holding in common only the fact that they are aggregates of objects.

Another factor, sometimes explicit, is the use of sociocultural anthropology as a model. The main impetus for higher-scale units of phenomena appears to be a desire to have units comparable to the "community" and whose classification will result in units analogous to "societies," "tribes," "cultures," or "peoples." In spite of this, prehistorians have long recognized that the resulting classes, such as phases, are not directly comparable to units in sociocultural anthropology, even if they have not always stated why. The difficulty in using a notion such as "community" for the scale of phenomena lies in the fact that communities' remains do not come in readily identifiable physical units. Communities must be inferred and thus cannot be the basis of distinguishing phenomena. The matter is further complicated because the objects which the prehistorian wishes to treat as an aggregate are situated in both time and space, rather than space alone, as is the case with most sociocultural units.

A final factor, perhaps as much a result as a cause, is that the devices used to create units at this scale are usually explicated as either grouping or taxonomic classification, neither of which lends itself to conveying the means by which decisions are made by the investigator. The tautological relationship expressed between classes at this scale and the phenomena certainly is a characteristic of these devices. Regardless of the rationale provided for unit construction at this higher scale, it is apparent from actual practice that classification, not grouping, is the means by which units are formulated, since the units have distributions, new information can be identified with previously established units, and even, in some cases, the necessary and sufficient conditions for membership are stated (e.g., determinants).

Admitting the desirability of a scale of phenomena larger than discrete object and recognizing that such units must be by necessity aggregates not as readily identifiable as discrete objects, it becomes necessary to state the characteristics that units at such a larger scale should display. It is not the purpose of this treatise to write anew the formal theory of prehistory, but simply to provide a framework for using what has been written. Nonetheless, at least a name for the units at this higher scale is required to continue any discussion, even if the unit cannot be precisely defined. Notions such as site (the place where the archaeologist digs) or component (which presumes the classifications for identification) will not suffice. The actual unit employed is the "collection." The object of classification is collections of discrete objects obtained in a spatially restricted area. How the space is restricted and the conditions its contents must meet is the focus of the problem. Judging from the literature, it is usually done intuitively. Yet there are clearly a set of goals which these collections, sometimes labeled assemblages, are intended to meet. First, it is evident that the objects making up the aggregate are intended to include only those made by the same set of people. Secondly, the set of objects is intended to represent those people at that place, that is, the collection or assemblage is to represent a sample of a spatial cluster. Thirdly, the set of objects is intended to represent a specifiable temporal segment, usually a period of continuous residence. In my own work the need for such units has arisen and the unit has been termed *"occupation,"* defined as a spatial cluster of discrete objects which can reasonably be assumed to be the product of a single group of people over that period of time during which they were in continuous residence at that particular locality. Quite obviously, the occupation is a tactical unit, not a theoretical one, and adapted to a specific body of data, in this case seasonal settlements. It is not generally useful. One need, for example, only consider the remains left by civilized peoples who may be in "continuous residence" at a given locality for a thousand years to appreciate the limitations. A tactical definition such as this does point toward a solution. The terms of the definition must be discrete objects—these are phenomenological and identifiable. The spatial boundaries will necessarily be based on proximity of discrete objects, again recognizable in phenomena. The spatial clusters of objects must be accountable as the products of a single group of people and deposited over a finite, specifiable time. A more workable definition might be constructed by treating the temporal element in terms of comparability and defining occupation as *a spatial cluster of discrete objects which can reasonably be assumed to be the product of a single group of people at that particular locality deposited over a period of continuous residence comparable to other such units in the same study.* This too is a tactical definition, not a theoretical one, but it does offer a more general solution than does the first, and effectively compresses the dimensions of time and space from the unit so that it is comparable to discrete objects. This kind of unit definition suffers from the principal disabilities of most archaeological notions; so defined, the units of one study are not comparable to those of another. Be this difficulty as it may, the term "occupation" can be used for the scale of phenomena above that of "discrete object" if cognizance is taken of the fact that the label only suffices to continue the discussion and does not constitute the resolution of this serious problem.

One thing ought to be clear. Whatever set of rules may be developed to distinguish the phenomena being treated as occupations, only a portion of remains treated as discrete objects can be classified at a higher scale, perhaps only a modest portion. Because the occupation, however defined, will always be an aggregate of objects lacking physical discreteness, it will be subject to alteration through time by simple mechanical motion greatly reducing the number of clusters which can be reasonably assumed to be the product of a single group of people or any other specified condition. This reduction in sufficiency is to be expected as a consequence of the greater precision and information required. It will always be the case that more archaeological remains can be accounted for and explained as discrete objects than as occupations or any other kind of aggregate.

Less inclusive scales than that of discrete objects present fewer difficulties than does the more inclusive scale, primarily because they are less frequently used and because they are component rather than composite elements and thus can make use of manipulatory discreteness for their identification. Less inclusive scales are always "pieces" or features of discrete objects—the problem of identification is simply a matter of conveying the manner in which discrete objects are to be divided. While not a common level at which paradigmatic classes are formed for the purpose of making hypotheses, the scale of "part artifact" or attribute is very familiar in the literature, for it is at this scale that features which are both the elements used in definition of classes and description of their *denotata* are formed. These are intuitive when used as the analytic units for classification at the scale of discrete objects; however, paradigmatic classes have been usefully formed at the scale of part-artifact.

Proceeding from the least inclusive or smallest scale to the most inclusive, those scales customarily used in prehistory are the "attribute" (of discrete object), "discrete object" (including both portable and non-portable objects), and "occupation" (aggregate of discrete objects). These scales constitute the three "sizes" of artifacts ordinarily treated by classification in the discipline. All three have the same properties of human involvement, and all are treated as things. They differ in physical size and the manner in which they are perceived, differences which profoundly affect their recovery as data but differences which do not enter into their properties as alternative units of classification. Obviously the inferences made about artifacts at each of the scales are widely different, and this is the reason for employing several rather than a single scale.

The spatial cluster that constitutes an occupation is in some senses empirically discrete—through time, with additional activity both natural and cultural, this discreteness is lost to a greater or lesser extent. Today additional scales are being recognized, at least experimentally, that lie between the discrete object and the occupation, clusterings of objects within occupations which give them their patterned character. Such treatments are not yet routinized to the extent that a single or series of intermediate scales are widely recognized in the fashion of attribute-object-occupation and thus are not properly treated here. Simply noting such a direction in prehistoric researches serves two purposes: (1) it emphasizes the arbitrary and customary nature of the three-scale system, and (2) points up the possibility of *extracting,* currently by means of distributions and associations of objects within occupations, phenomena at scales not ordinarily perceived as such. All of us would see both objects and occupations and things; not many would perceive an activity locus as a thing, yet our "common sense" perception is no measure of utility, even though the three-scale system is just such "common-sense" perception.

At any given scale an infinite number of classifications is possible, with alternative classifications for the same objects. Different classifications may have different purposes and thus make use of different criteria. Such alternative classifications often differ in level. Taking again the type-variety system, "wares," "types," and "varieties" are alternative classifications of potsherds, three classifications differing in level but treating the same scale of phenomena. Such ranked constructions must not be confused with classifications of different scales such as the "mode," "type," and "phase" classification presented in the following pages. Further, in constructing classes two scales must always be used. The features employed as criteria will be drawn from a scale below that of the classes. To formulate classes of discrete objects, features must be drawn from the scale of attribute. Likewise, features defining classes of occupations will be drawn from the scale of discrete object or attributes of discrete object, or both.

The following section identifies the specific classifications currently employed in prehistory in terms of the framework just set forth. Perhaps as much as ninety per cent of all classification used in prehistory, when sufficient information is provided, can be treated as members of this system. This is, of course, in spite of divergent terminology in which different units are called by the same label (as is the case with type) and the same unit is labeled with different names (as is the case with mode), and in spite of a lack of a precise separation between the classes and their *denotata* and the inconsistencies introduced by this failure.

Classification in Prehistory

Figure 15 presents the set of classificatory units widely employed today, using the most common terms for the units in-

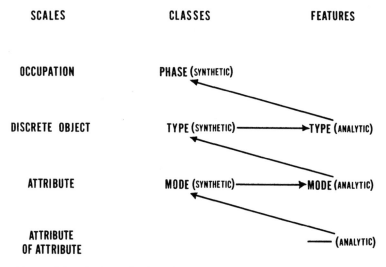

Figure 15. Scales of classification customarily employed in prehistory.

volved. It is important to note that at each of the customarily chosen scales two different kinds of units occur, one a synthetic or classificatory unit and the other an analytic unit. The unit mode, for example, is identical or can be identical in content when used in the definition of type or as a paradigmatic class in its own right. The distinction between analysis and synthesis is relative. If modes are used as features in the definition of types, they will be treated *as if* they are intuitive classes of attributes even if they themselves are the product of an explicit classification at the scale of attribute. This is so because paradigmatic classification presents only one set of definitions: the features used to phrase the definitions are defined outside that particular classification.

Beginning at the lowest scale of phenomena, Figure 15 indicates an unnamed analytic unit used to define modes, the units at the next highest scale. This unit is implicit in the literature, principally because modes are usually considered "indivisible units," the smallest possible qualities, a view which obviates an explicit statement of definition. The inclusion of such a unit at a scale beneath those ordinarily employed serves mainly to allow for the definition of modes, not because it is frequently encountered. The unnamed unit in Figure 15 can be defined as *an intuitive cultural class of attributes of attributes of discrete objects*. Intuitive, in this and the following definitions, indicates that the unit so characterized is not the product of an explicit classification in the particular context employed. Cultural should be understood as meaning that the elements of the definition, be they intuitive or explicit, can be assumed to be the product of human activity, that is, artificial. Insofar as I am aware, there is no synthetic or classificatory unit at the scale of attribute of attribute of discrete object.

"Mode" is the term applied to classes, both analytic and synthetic, at the scale of attribute of discrete object. This classificatory unit plays the crucial role in the system of classifications employed in prehistory. Some investigations are conducted at this scale (e.g., Rands cited in bibliography), and so modes sometimes are defined classificatory units. So employed, mode is defined as *a cultural paradigmatic class of attributes of discrete objects*. This, however, is not the most important or frequent use of this class. Its most important use has been as the analytic step providing definitions for classes at the scale of discrete object. Types are defined and described in terms of modes. This unit is probably the subject of more terminological abuse than any other. Modes have been and are called "features" (meaning characteristics), "attributes," "factemes," and "traits," to name only a few. Employed in an analytic context, mode is defined as *an intuitive cultural class of attributes of discrete objects.* Since the mode is the smallest-scale unit commonly employed in prehistory, it bears the burden of converting classification in general into cultural classification for prehistory. It is usually here that the assumption that *attributes which are the product of human activity and which recur or are shared may be treated as the product of shared ideas* is injected into the work of the prehistorian. This assumption itself is sufficient for the creation of modes as analytic units. Unfortunately, modes, irrespective of what they are called in a given piece of literature, are frequently dealt with only incidentally. The comparative work required in assuming a given attribute or set of attributes to be the product of human activity is not frequently presented in explicit form. Attributes are not cultural; they are part of the natural world. To assume that a given set of attributes is the product of human activity requires a comparative study. Modes treated under the terms "trait" or "characteristic" seem particularly plagued with this lack of serious concern. What is cultural varies from place to place and from time to time. The mode functions in prehistory to isolate sets of attributes which are cultural in a particular context. Undoubtedly because of the sloppy treatment this matter has received in many cases, modes that would have been useful for the purposes of the given study have been left unused while other "attributes" or "traits" employed are not even cultural, let alone relevant to the problem considered. The importance of modes in prehistory cannot be overemphasized. They themselves are sometimes used to provide the basis of hypotheses and inferences about styles and technology as well as chronological problems, but their most important use is in the definition of all further cultural classes employed in prehistory. The term mode has been chosen from the plethora of terms because of its chronological priority in association with good definition (defined in Rouse, 1939).

The majority of investigations in prehistory are conducted at the level of discrete objects if only because of the ease of identifying this scale. As is the case with mode, units at this scale are employed both as units of study and as means of defining units at still higher scales. In contrast with mode, how-. ever, the unit at this scale, type, is most frequently used as a unit of study rather than an element for definition. In this synthetic context, type is defined as *a paradigmatic class of discrete objects defined by modes*. It is possible here and at this scale to specify the units used for definition, in this case modes, and thus drop the cultural adjective for type. Types must be cultural if they are defined by modes. This also obviates any need to make the basic assumption more than once. Types are sometimes used to define units at the next higher scale. In this context of analysis, type is defined as *an intuitive cultural class of discrete objects*. While types so used are intuitive at the next scale of classification, in practice they almost never are, for they have been formulated as units of study defined in terms of modes and *then* used as elements of definition at a higher scale.

Like mode, type has seen considerable terminological abuse, more in the direction of different kinds of units being called types than in different names being used for the unit here called type. "Type," especially when qualified as "descriptive," is often used for intuitive groups which do not in any respect meet the criteria of classes and is thus employed as a synonym for English "kind." Type is also applied to the products of grouping devices, particularly statistical clustering, and this is the most serious terminological problem, given the magnitude of the distinction between groups and classes. On the other side of the coin, the terms "variety," "ware," "style," and "functional class" are but a few of the names occasionally applied to paradigmatic classes at the scale of discrete objects. Most of these terms reflect not the kind of unit, but the particular purpose for which the class has been constructed. Thus functional classes are usually types which are explicitly created for the purpose of inferring the function of discrete objects. The terms "ware," "type," and "variety" in the type–variety system name types that differ in level: "wares" being the types which are used for comparisons over large amounts of time and space, "types" being the types used for comparisons within small areas and limited amounts of time, and "varieties" being types used primarily for intra-site comparison. As has been pointed out, all three are paradigmatic classes, or can be, for discrete objects differing in level. The choice of definitive modes is predicated on the purpose to which the units are to be put.

To reiterate: Types are paradigmatic classes of discrete objects defined by modes. Types are not groups of objects, but classes whose *significata* consist of sets of modes stating the necessary and sufficient conditions of membership. Since these conditions are modes and modes are cultural, types are cultural.

There are substantial difficulties in identifying the phenomenological units at the next highest scale, that of occupation, and thus it is not surprising that there is considerable confusion (both conceptual and terminological) about classification at that scale. The most commonly employed term for these classes is "phase"; however, the theoretical rationale for the construction of phases is usually phrased as a kind of numerical taxonomy. This particular rationale is considered in the following chapter. It is sufficient here to note that the units formulated have all the characteristics of paradigmatic classes (e.g., distributions in time and space, plus unranked or unweighted definitive criteria called determinants), and that they can be used to identify new data. Only classificatory or synthetic units appear to be constructed at this scale. Phases do not serve as analytic units for any higher scale of phenomena. In spite of divergent explanations for the *phase*, it is employed as *a paradigmatic class of occupations defined by types and/or modes*. Phases are identified as recurrent sets of types or, less frequently, modes. In the literature, "phase," "focus," and "culture" are often used interchangeably for paradigmatic classes of occupations. The terminological difficulties are increased by the use of such labels as "complex," "industry," and "assemblage" to refer to both the *denotata* and the *significata* of the classes. The term "component" has seen fairly consistent usage as a label for the *denotata* of a given phase at a given locality.

The construction of phases in the discipline has largely been directed toward the construction of classes which can be called "whole cultural," that is, classes which link together the various remains of a single set of people. It has been customary to call paradigmatic classes of occupations other names when constructed for purposes other than "whole cultural" units. Many of the "larger units" considered later in this chapter are phases; that is, they are paradigmatic classes of occupations, but they are not necessarily "whole cultural" units. The definition of phase presented here is not restricted to classes for any particular problem. There may be, and indeed are, phases formulated on the basis of functional criteria as well as those formulated along the more customary lines with stylistic criteria.

In summary, there are three fundamental scales at which paradigmatic classes are formed in prehistory: (1) attribute of discrete object, with the resulting classes termed modes; (2) discrete object, with the resulting classes termed types; and (3) occupations or aggregates of discrete objects, with the resulting classes termed phases. There is implicit a fourth scale, that of attribute of attribute of discrete object, at which the units are not named and which function only as the analysis for modes when such is attempted. Modes are basic to the system because it is here that classification usually begins and the assumption which makes classifications cultural is employed. Modes serve both analytic and synthetic functions with the analytic function dominating. Types are the most widely used classes, almost always serving as synthetic units which in turn are used as analytic units. Phases are the highest scale of classes commonly employed, and they function entirely as synthetic units. Since types are defined in terms of modes, their *significata* being combinations of modes, types are cultural by definition. Phases can draw upon either modes or types for definition, and likewise are thus cultural.

Some Still Larger Units in Prehistory

It is the contention here that there are but these three scales at which synthetic units are ordinarily formed and a fourth which currently serves only as an analytic step leading to the definition of modes. There are, however, a number of named units in the literature which superficially appear "larger." Because of this quality of "largeness," there is no confusion in the literature about their nature as classes. The *denotata* are simply too numerous and too extensive to be assembled into a group, effectively preventing the confusion of class and *denotata*. These "larger" classes differ in no fundamental respect from those already discussed. They represent the very same classes (modes, types, and phases) but are defined for special purposes or at a level higher than that usually associated with classes labeled modes, types, and phases. Since there are a large number of such named units it is not possible or profitable to consider them all. The most widely used are tradition, horizon–style, horizon, series, and stage. The treatment of these notions here is brief, serving only as a pattern for how such classes may in general be regarded. Tradition, horizon, and horizon–style may be examined together since they are labels for "special cases" of the units just considered. These three units do not specify any particular scale, but rather are modes, types, and phases whose *denotata* display special temporal–spatial distributions.

Traditions are modes, types, or phases whose *denotata* display an extensive distribution through the dimension of time in conjunction with a limited distribution in space. The term tradition serves simply to name those modes, types, and phases with this kind of distribution. This particular distribution is the source of many inferences in prehistory concerning development, continuity, and "genetic relationship," and thus the need for a term to designate classes appropriate to such operations. Further, many explanatory models operate only within the confines of such classes, providing another important reason for their delineation. Frequently, traditions and one or more sets of other classes will be superimposed to provide the basis for inferring complicated temporal–spatial relationships. Classes which have the distribution of tradition are often defined upon functionally relevant features since such features tend to change more slowly than, for example, features of style.

Horizon and horizon–style are parallel constructions which designate classes whose *denotata* have extensive distributions in space coupled with restricted distributions in time. Horizon–style is most frequently applied at the scale of attribute, whereas horizon is the term used at larger scales. Again, the terms serve to designate classes with distributions of particular interest to many prehistorians, for the particular distribution labeled horizon or horizon–style serves as the basis for inferring such things as migration, diffusion, and contact.

Series and stages differ from tradition and horizon in that they do not serve to label classes of particular distributional characteristics. In the case of both stage and series, the level of classification is higher than ordinarily used, and the names serve to designate this change in level. Series and stages are usually, but not necessarily, at the level of phase. In both cases the defining criteria are relatively few compared to usual classes, with the result that their *denotata* are of wider occurrence in time and space, and they serve to link other classifications through coordinate *denotata.* Series are usually defined upon stylistic features; stages are usually defined upon technological features. Thus series tend to have coherent distributions in both time and space, whereas stages tend to have coherent distributions only in time. Because they involve few criteria, the amount of information provided by such classifications is relatively limited, and their main use lies in continental summaries and literature intended for lay consumption or introductory texts.

Various combinations of these larger units occur in the literature or are possible, especially if they are employed at different scales. The area–cotradition is an example of both tradition and horizon distributions used together. The more criteria that are employed, however, the more restricted the use of the resulting units. The important thing to recognize is that these grand classes differ in level and purpose but not in scale from the units considered here. Traditions are classes for the same scales as modes, types, and phases and are best treated as special kinds of modes, types, and phases. Series and stages likewise are classes for occupations (primarily) and thus are best considered phases defined by a small number of specially selected features.

Problem and Evaluation

The absence of an identifiable phenomenological unit above the scale of discrete object may be the most serious conceptual void in prehistory's formal theory, but by far the most serious operational difficulty is the chronic lack of problem and consequent lack of rational means of evaluating classifications. Thus, in turning to consider evaluation and problem, we are turning to classifications rather than the process itself. This difficulty is linked with, and perhaps in part is a result of, the confusion between the *denotata* of classes and the classes themselves, and concomitantly to the confusion of description (of *denotata*) with definition (of classes). A class "means" its definition or *significatum.* If, for example, we have a class defined red-rough-solid, the distribution of this class's *denotata* is that of only the objects as red-rough-solids *and nothing else.* Any hypothesis made to account for the distribution is an account of the objects as red-rough-solids. This class could not be used as the basis for inferences about shape, size, or any other characteristics of the objects identified as *denotata,* for these other characteristics are variable. Similar arguments could be made for association of *denotata* of different classes. The use to which a class may be put is a direct function of how it is defined. Problem and class definition are intimately linked.

As we have seen, definition of classes, regardless of the kind of classification, involves the selection of some classes of attributes as criteria. Thus the point at which problem enters classification is in the selection of definitive characteristics. A survey of archaeological literature shows three alternative treatments. Most commonly, the selection of criteria and the definition of problem is simply ignored. Classes are formulated by means unknown to the reader and perhaps to the formulator, and thus do not have an explicit *significatum.* The classes mean nothing and can legitimately be used for nothing. These cases may usually be recognized by the use of such terms as "descriptive," "inherent," "essence," or "natural." "Description" is usually proffered as the purpose. If, however, description is a purpose or problem, then any set of criteria will serve for all that is required is a set of words. There is no way to evaluate such constructions, nor do they have any meaning. They are natural, inherent, and represent the essence of the real world.

A second less frequently realized alternative is the explicit statement of the criteria chosen for the definition of classes but with no specified problem for which the classes are to serve. In this case it is possible to treat the classes as meaningful and to make hypotheses about their distribution and association, but there is no way to evaluate their utility. The criteria, while explicit, are commonly drawn at random, and the classes are not a useful organization for any problem. Indeed, this alternative seems to be realized when the object is "description," and the classes are not constructed for any use beyond a device to say what was found where and to provide terms for the ubiquitous "site-to-site comparisons." All the comparisons mean, however, is that thus and such types are found in thus and such places, in spite of the speculation sometimes associated with such "comparison."

The third alternative, the statement of both problem and definition of classes, is the least frequently realized. The statement of a problem for which the classification is to serve as the organizing device provides the rationale for making the choice, be it overt or covert, that must be made in defining classes. The utility of a classification then becomes testable. Either the classification will organize data for, say, a chronology, or it will not. The particular choices made can be weighed against other possible choices and those best suited to the problem selected. While implicit in many important respects, James Ford's pottery classifications for the Southeastern United States are some of the best examples of problem-oriented classification. His sole concern was classifications for ceramics which could be used in constructing chronologies with the seriation method. While it is not often possible to separate the *significata* of his types from the description of the material assigned to them, his own general statements indicate how the decisions were made: only those combinations of modes which had short distributions in time were suitable. His definitions are stylistic. Further, he admits the possibility of making wrong selections which will not prove useful for his purposes and which will have to be "reformulated." While it is possible to recognize Ford's problem and to state generally how he employed classification, principally types, for its solution, his chronic failure to differentiate type definitions from the description of their *denotata* makes for difficulties in using his material as an example.

By way of summarizing this third alternative it is useful to introduce an example which begins with the selection of criteria for the definition of types and follows through to their evaluation. For these purposes the problem can be stated as chronology, the method for which the classes must function as seriation, thus closely following Ford's interests, and hopefully elucidating some of the operations which make it work. Let us say we have a series of pottery collections from the set of localities shown in Figure 16. Our immediate purpose will be the the selection of a series of dimensions of modes suitable for seriation—modes whose primary variation in representation in the area of concern is through time rather than through other di-

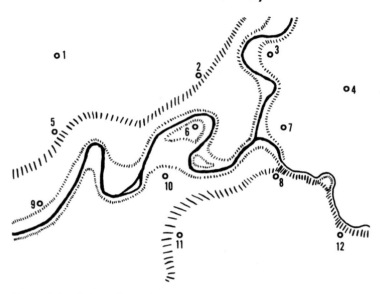

Figure 16. A hypothetical region showing the location of sites and distribution of modes. The modes occur as listed below and are abbreviated: cordmarked, cm; plain, pl; decorated, d; undecorated, **d;** shell tempering, s; limestone tempering, l; and other stone tempering, o.

1. cm, pl, d, **d,** s, l. 2. cm, **d,** o. 3. cm, pl, **d,** l, o. 4. cm, pl, d, **d,** s, l. 5. cm, pl, d, **d,** s. 6. cm, **d,** o. 7. cm, **d,** l, o. 8. cm, **d,** l, o. 9. cm, pl, d, **d,** l. 10. cm, pl, **d,** l, o. 11. cm, pl, d, **d,** s, l. 12. cm, pl, d, **d,** s.

mensions (e.g., space). Seriation orders groups by arranging them so that the distribution of the *denotata* of historical classes is continuous and if the frequency of occurrence is treated these frequencies take the form of a unimodal curve. For the purposes of illustration we need consider only the first model, that of continuous distribution, usually termed occurrence seriation.

One might begin simply by combining all the collections and distinguishing various features of their construction, decoration, and the like, being careful to ascertain their artificial nature. Since styles are desired, certain kinds of attributes will intuitively be important from the beginning, such as decoration. Other kinds of attributes, such as shape, may have strong functional components; and still others, such as clay, spatial components. These problematic features will greatly outweigh those which can reasonably be assumed to be relevant. From these will have to be distinguished features useful for defining historical types. As with anything, initially one must guess as to which will be useful and which will not. The guesses will be phrased as hypotheses that x mode is historical in its distributional characteristics. Various means are available to enable one to make relatively good guesses. For example, having no-

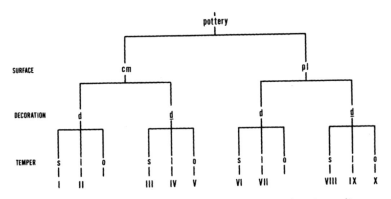

Figure 17. A paradigmatic classification utilizing the three dimensions of modes plotted in Figure 16. Abbreviations for modes are those used in Figure 16.

ticed what features occur at what locations, one could plot the spatial distribution of the modes as is done in Figure 17. Thus controlling one dimension of variation one can narrow the field of choices by reasoning:

    1. Modes which occur at only one location are useless since they do not provide a means for comparing the various collections.

    2. Modes which occur at all locations are not likely to be useful since they change too slowly to provide precise comparisons.

    3. Modes that exhibit distributions closely linked to (a)

geography or (b) environments are obviously variable in terms of space and/or function to a significant degree and are thus unsuitable.

The search can be narrowed thus to modes which occur at several but not all locations and which do not exhibit any clear-cut patterning in space or correlation with environments. The justification for such choices could take the form: If cultural attributes have been chosen, they can be expected to have a patterned distribution. Features which display a random distribution in space must be variable in uncontrolled dimensions —among others, time. In Figure 16, modes in the dimensions of temper, decoration, and surface finish have the desired distribution, whereas the dimensions of shape and color as well as clay would appear to be patterned in space or correlated with environment. Limiting the initial choices in such a manner gives one reason to believe that types defined by these modes will be worth testing to see whether or not they are in fact historical. Noting this kind of distribution does not mean *ipso facto* that the unpatterned sets of modes will define useful historical types, for there are many other possible explanations for the lack of spatial pattern.

Figure 17 shows a paradigmatic classification utilizing three dimensions of modes: surface treatment divided into modes "cord-marked" and "plain"; decoration divided into modes "decorated" and "undecorated"; and temper divided into modes "shell," "limestone," and "other stone." Two of the twelve classes so generated have no *denotata*, that is, no sherds are cord-marked, decorated, and tempered with stone other than limestone; and no sherds are plain, decorated, and tempered with stone other than limestone. All the other classes are given names, Types 1–10. The next step will be identifying each location in terms of the types represented in its collection.

The final step is the seriation, the arranging of the Groups A–L so that the distribution of Types 1–10 is continuous. The seriation actually constitutes a test of the hypotheses made in selecting the definitive modes. If the groups can be arranged so that all of the types display continuous distributions (Figure 18), then the selection hypotheses can be considered correct. As anyone who has frequently employed seriation is aware, randomly devised classes will not closely approximate the required distribution. If the groups cannot be so arranged, with appropriate allowances made for the effects of sampling error upon the representation of the types, then the hypotheses made in the selection of definitive modes is shown to be incorrect and the types must be rejected. There may be a number of reasons why a set of groups cannot be seriated, aside from applying the technique to data for which it is inappropriate. The dimensions chosen may be appropriate (e.g., tempering is historical), but the divisions into modes incorrect (e.g., shell, stone, and sand

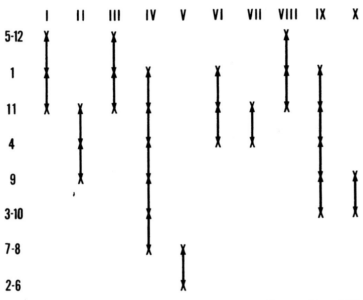

Figure 18. An occurrence seriation of the data in Figures 16 and 17. Localities 5 and 12, 3 and 10, 7 and 8, and 2 and 6 cannot de differentiated on the basis of presence and absence alone.

rather than shell, limestone, and other stone); or the dimensions may vary significantly in dimensions other than time, thus tending to randomize temporal variation. There are means available to solve for these possibilities, but these are beyond the scope of this illustration.

To be certain that the order produced by a seriation is a chronology will require additional seriations of the same set of groups in terms of other materials (e.g., projectile point types, house types, burial types, etc.), and only that order which is repeated from one seriation to the next can be treated as a chronology. Insofar as testing the utility of a set of classes is concerned, however, the ability to seriate the groups suffices.

Without a specifically stated problem there is no way, even if the definitive criteria are explicit, to justify the selection made. If the problem is specified and the *significatum* explicit, then: (1) the relevance of the criteria chosen to the problem is testable, that is, an assessment of utility is possible; and (2) given alternative classifications, the most sufficient and parsimonious can be chosen.

As it stands, however, most classifications are taken for granted. There is little or no concern with how the classes came to be and why. Classification is very often done for its own sake, and this requires no evaluation or concern. A prime contributor is the feeling that all of us have that *a* thing should have *a* name. The only problem is deciding what name to use. The implication from our discussion is that there will be as many classifications as there are problems. This is certainly not new, for just such an assertion is the crux of J. O. Brew's arguments cited earlier. His admonishments have not been generally heeded, because it is discomforting to the archaeologist to have one and the same set of artifacts belonging to ten different phases, or the same object assigned to ten different types. Nonetheless, the classes used must be a function of the problem if they are to mean anything—if they are to be subject to testing and evaluation and if they are to be accepted because of utility rather than on faith.

7

GROUPING IN
PREHISTORY

Groups, aggregates of phe-
nomena, are the focus of scientific study, for it is phenomena
that science seeks to explain. However, it has been argued that,
as the device for construction of groups, grouping is entirely
inappropriate to scientific endeavor, and that the only groups
profitably employed are the *denotata* of classes, especially the
denotata of paradigmatic classes. To briefly review the rationale
for the exclusion of grouping as a device for scientific unit construction: (1) groups constructed by means of grouping devices can have only extensional definitions consisting of a list
of members; and (2) thus such units cannot recur through
time and space (a requirement of prediction and control) or
be shared (the special requirement of the notion culture). Because groups so constructed consist *only* of their members,
they are applicable only to the members originally included and
cannot incorporate new information. Such groups are history-
bound, inappropriate to, and indeed impossible to use for
measuring change in either time or space.

The cursory considerations undertaken in this chapter,
then, are germane to our purpose only insofar as grouping has
been used or suggested in prehistory to create units. The
major aim is to identify grouping so that it can be avoided. The
identification of groups which are the products of grouping as

opposed to groups which are the *denotata* of classes would be an easy matter were it not for a penchant of prehistory's literature to present a description of *denotata* without presenting the classification by means of which the *denotata* were assembled. This procedural error makes it difficult to distinguish grouping and classification in the literature, for most of the readily usable criteria are not presented (e.g., explicit definitions). Adding to this difficulty is the lack of problem in many studies. When units are constructed for their own sake or when the "problem" is "description," the units are not used beyond their names, eliminating any possibility of identifying the nature of the unit from its characteristics of use. This is most unfortunate, for if a unit is employed for some purpose, the use will suffice to distinguish between those units which are the product of grouping and those which are the product of classification. These circumstances are sufficiently common that most units used by prehistorians are amenable to interpretation as either groups or classes. Only when the entire discipline is considered is it possible to assess the nature of the units commonly employed.

Admitting the difficulty of distinguishing grouping from classification in prehistory as a function of the sloppy treatment accorded systematics, all that can be done is point out some of the more frank uses of grouping and the problems which result from these attempts and proposed procedures. From the outset it should be evident that any method, irrespective of its pragmatic utility, can be constructed on paper. The only requirements it must meet are those of logical consistency. Thus it is possible, and, in fact, occurs, that the rationale for some specific study's units may be presented as one or another grouping device, even when the actual procedure has been paradigmatic or taxonomic classification and when the device offered as the rationale could not conceivably have produced the units attributed to it.

Insofar as I am aware, grouping devices have been used as the rationale or proposed as the means of unit construction only at the scales of discrete object and occupation. Both numerical taxonomy and statistical clustering are in evidence for discrete objects, while only numerical taxonomy has been used for occupations. In all cases the units have been labeled with terms used to designate classes so that "type" in the literature can mean either units which are *denotata* of paradigmatic classes *or* the products of grouping. The remainder of this chapter will attempt to show how grouping has been used, what the characteristics of its use are, and the problems which result.

Statistical Clustering

In terms of method, there is nothing which can be added beyond what has already been presented in Part I, since that discussion is based largely upon the use of statistical clustering in prehistory. The primary advocate of statistical clustering in prehistory has been A. C. Spaulding, who first detailed the approach in his 1953 "Statistical Techniques for the Discovery of Artifact Types." The approach begins with a paradigmatic classification. Indeed, Spaulding presents the clearest statement of paradigmatic classification that can be found in the prehistoric literature, being particularly noteworthy in the clear recognition of the dimensional character of the defining modes. The frequency of the definitive modes is tabulated for the collection being considered, and an expected frequency of combinations of the modes in discrete objects assuming a random association of the modes is calculated. Essentially this is a statement or prediction of the number of combinations that will be found strictly as a function of the frequency of the modes. The next step is the tabulation of the actual combinations of modes found in the collection, and the results of this tabulation are compared with the expected frequencies. The outcome of this comparison, which takes into account the size of the sample considered, is the isolation of combinations of modes which cannot be accounted for as the result of random association and vagaries of sample size. There are, of course, two possible kinds of clusters: negative ones, combinations which do not occur or which occur much less frequently than would be predicted on the basis of random association; and positive ones, clusters which occur more frequently than could be predicted on the basis of frequency of the individual modes. The ability to detail what combinations are actually realized out of those combinations that are logically possible is one of the distinct advantages of explicit paradigmatic classification over other kinds of arrangement. It provides immediate feedback in the form of a non-random distribution that the attributes chosen are the products of patterned behavior. Should the distribution be random, it is reasonable to assume that the attributes chosen are not culturally significant in the form in which they have been conceived. The isolation of positive clusters is taken to be a discovery of genuine tendencies on the part of the makers to combine sets of attributes, and the positive clusters labeled types, or rather, potential types. They are potential only, because if two or more significant clusters differ in a few modes (i.e., are closely "similar"), they will be grouped together as a single type of two varieties.

Up to this point in the procedures, there are no serious difficulties. Two sets of classes are in evidence, the modes used to characterize the material and their combinations into paradigmatic classes (Spaulding's attribute combinations). The comparison of the frequency of the modes with the frequency of their combination indicates that the choices of modes are culturally significant. The difficulty arises when those combinations which are heavily represented are singled out as "types," something quite different from the sense in which type is usually employed, for here the types are directly linked through the counts made of attributes and combinations to a particular body of artifacts. Further, not all the objects in the collection need fall in positive clusters, and those which are infrequently represented are not recognized as types but relegated to the status of "abnormal" combinations of modes. It is likewise entirely possible that no clusters, either positive or negative, might be found, and thus the collection be regarded as having no types or as being all of one type.

What has been done is clear, as is the nature of the units which result from this approach. The *denotata* of paradigmatic classes (termed in the approach "attribute combinations") at a given location in time and space have been counted and this tabulation compared with a tabulation of the *denotata* of the definitive modes (termed in the approach "attributes"). The comparison of these two sets of *denotata* differing in scale is then used to create units called "types." The "types" are quite obviously groups of real objects. Any kind of counting requires phenomena, and any kind of units based on count in any fashion are phenomenological, that is, groups. This situation could be treated as a particular case of the general confusion of classes with their *denotata* in prehistory, the name "type" simply being applied to the objects assigned to the type at a given locality, were it not for the lumping of closely "similar" clusters into the same unit as varieties.

Further difficulties arise when the infrequently noted combinations are regarded as abnormal combinations of modes. A combination of modes which is infrequently represented at one locality and point in time, and thus an "abnormal combination," will usually be in some other locality and time frequently represented and thus at that locality a "type." Popularity varies through time and space, and units based upon popularity necessarily vary as well. The peculiar consequence of employing statistical clustering is the creation of sets of units unique to each sample location—giving rise to a "rubber yardstick." Being bound to the occurrence of attribute combinations at specific localities, the meaning of the units will change with the frequency of representation. Types so constructed cannot provide means of either comparing localities with one another or measuring formal change. In short, the units are descriptive and are not capable of providing the terms for explanation. Nor, in the absence of problem, are they testable. Figure 19 presents a comparison of the distribution of *denotata* of paradigmatic classes labeled Type A through Type D with statistical clusters labeled "Type a-Type f." In this simplified hypothetical case, the vertical axis of the diagram represents time, the width of the curves the frequency of occurrence of *denotata* of paradigmatic classes. The paradigmatic classes do not change through time but rather the frequency of occurrence or the presence and absence of their *denotata* change. All of the localities represented by the bars in the diagram can thus be compared with each other, the paradigmatic classes providing the basis of comparison. The statistical clusters, it will be immediately noted, are restricted to specific localities, being actual groups of artifacts, and thus these units themselves change through time and provide no basis for comparing the various localities. In situations requiring larger numbers of types, the contrast between clustering techniques for unit construction and paradigmatic classes would be even more dramatic, though more complicated in its portrayal. The addition of new localities, new data, will result in a proliferation of the number of clusters but will not affect the number of paradigmatic types.

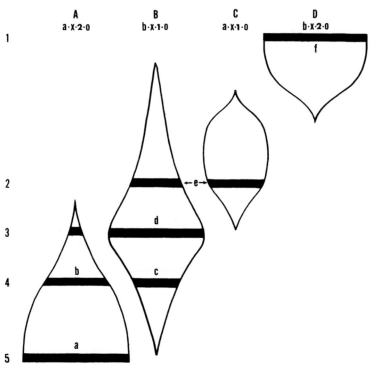

Figure 19. A comparison of statistical clusters (a-f) and the distribution of paradigmatic classes (A-D). Localities (1–5) are represented by horizontal sets of bars; the frequency of occurrence of **denotata** is represented by horizontal sets of bars; the frequency of occurrence of **denotata** is represented by the width of the curve. The definition of each paradigmatic class is shown immediately beneath the type designation.

These difficulties were recognized by Spaulding in proposing the device, for he clearly states that the "types" are groups of objects, and that, further, the "types" are restricted to single localities and single occupations of these localities, the essential identifying characteristics of grouping devices. However, the importance of these difficulties, seeing them as difficulties rather than characteristics, is not appreciated, for the "problem" for which the statistical clusters were to provide an order was "description." Thus, in the absence of any testable, definable goal, there was no means to judge the utility of the results. Any kind of units which provide a means of naming will suffice "description." No complicated devices are required, nor do they have any demonstrable advantages, save maybe intellectual satisfaction, over any other means of naming.

The publication of this device for unit construction led to a long and rather involved argument in the literature with James Ford, who was using types in the sense used herein (paradigmatic classes). The major components of this argument are listed in the bibliography, and make most useful reading, clearly demonstrating the problems presented by using the term "type" for widely different kinds of units. A careful reading of this argument will also demonstrate the utility of making a distinction between groups and classes in attempting to understand the archaeological literature. As in all cases in which argument is more about words (in this case "type") than substance, the argument slowly dies instead of being concluded decisively. It is worth pointing out, however, that the statistical cluster has not seen use in prehistory for any operations beyond those kinds of studies in which creating units for their own sake—description—is the goal. Spaulding's clear exposition, some of the finest in the archaeological literature, is often cited as the rationale for "type" even when what is actually done is paradigmatic classification such as Ford argued for, though hardly as succinctly.

The discussion of clustering leads rather directly to a larger problem, the quest for "folk classifications." Presumably this quest is a motivation behind statistical clustering as a means of unit formation, given that one of its stated aims is the discovery of genuine tendencies on the part of the makers to combine modes. Aside from the fact that there is no way to know whether or not the modes initially used were recognized in some cognitive sense by the makers, the irrelevance, indeed the detriment of such "folk classifications" to scientific investigation, has already been argued. It must be emphasized that employing paradigmatic classes in no way prohibits a statement of these combinational tendencies. These variable representations of combinations are, however, statements about the distribution of *denotata*, not characteristics of the classes.

For example, in Figure 19 one may by inspection or by the methods outlined as statistical clustering characterize the time and space represented at Locality 5 by the tendency for a single combination of modes (a–X–2–o), Locality 2 by the tendency for two combinations (a–X–1–o and b–X–1–o) which differ in a single mode, and so on, without binding the analytic units to the circumstances that obtain at any one of these localities. One might further *speculate* that the people involved in Locality 5 recognized but one type; that those at Locality 4, one type of two varieties; and so forth. Insofar as there are no means available to test these statements they must remain speculations. Folk classes constitute interesting data, artifacts, when and if they can be recovered. They are to be explained; they are not an explanation. The aim of making analytic categories coincident with folk categories quite obviously will always result in the units being groups, since the categories themselves are phenomena. This particular goal evidenced in some archaeological studies is an excellent case in point with regard to the inappropriateness of sociocultural anthropology, from which the notion derives, as a model for prehistoric investigation.

Numerical Taxonomy

Numerical taxonomy has been proposed as a device for creating units at the scale of discrete objects; however, this is not yet widely practiced. Numerical taxonomy produces groups, and thus the units have the same characteristics as statistical clusters insofar as their utility in scientific endeavor is concerned. They are contingency-bound, undefined and undefinable, and restricted to the material from which they are derived. They cannot serve as the basis for comparison, nor can they incorporate new data without changing the structure of the units. Such groups bear the same relationship to the distribution of paradigmatic classes· as do the statistical clusters in Figure 19. The advocacy of numerical taxonomy as a means of unit construction at the scale of discrete objects has followed the systematic exposition of this device in the biological sciences and incorporates the statistical sophistication characteristic of these disciplines.

Far more important than the proposed use of numerical

taxonomy at the scale of discrete objects is the widespread use, or at least advocacy, of numerical taxonomy to construct units for aggregates of discrete objects, the scale of phenomena herein called occupations. This use of numerical taxonomy long antedates the appearance of this device in the biological sciences and appears during the 1930's in a non-statistical form. In fact, it is almost the only device explicated in the prehistoric literature for unit construction at the scale of occupation, and this in spite of the fact that the units actually employed are, when identifiable, almost invariably paradigmatic classes.

The general approach is best stated in its early form by William McKern, one of the inventors of the device, in "The Midwestern Taxonomic Method as an Aid to Archaeological Culture Study," published in 1939. There it is proposed that aggregates of discrete objects, collections which are termed components, be compared with one another in terms of "traits" in order to assess the degree of similarity exhibited between collections. No formal coefficient of similarity or agreement is employed, but, rather, the expression of similarity takes the form of a list of linked (shared) traits and diagnostic (unshared) traits. The linked traits, of course, are the ones used to create the units while the diagnostic ones are to serve the purpose of identification. It is apparent even from the outset that grouping, in this case numerical taxonomy, and classification are undifferentiated in the system, the linked traits clearly belonging to a grouping device, while the diagnostic traits suggest that the groups are to be employed as classes. The lowest-level unit is the component which is considered empirical, that is, part of the phenomenological realm and the referent for the other units in the system. These components are successively grouped on the basis of similarity into foci, aspects, phases, patterns, and bases, with foci being the most similar units, the bases the least similar. It is further observed that styles are linked traits between foci and that as one goes to higher levels the linked traits progressively become first more technological and then more functional. This generalization, which amounts to saying that styles have smaller distributions than do technologies or functions, admits the possibility of viewing the Midwestern system as a series of classifications, each level being defined by different kinds of criteria. This impression is, however, most superficial. Higher-level units effectively group lower-level units; the components assigned to Focus 1 will not be split among two or three aspects but will belong to the same aspect. The only way in which this coordination of units at different levels may be achieved is the inclusion of all the criteria at the lowest level (focus) and reducing the number to

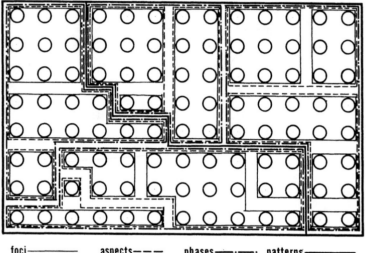

foci———— aspects— — — phases—·——· patterns————

Figure 20. A schematic diagram showing the coordination of various levels within the Midwestern Taxonomic System. Components are represented by the circles.

derive the next level, and so forth. Not only are styles linked traits at the level of focus, but so are all the other traits which are linked at higher levels. Thus the difference, for example, between foci and patterns is not in the kind of criteria, but in the *number* of criteria, which are held in common. This characteristic of coordination of various levels in the Midwestern Taxonomic System is illustrated in Figure 20, in which components are represented by small circles and the various groupings by rectangular boxes. All of the boxes include other boxes and none of them intersect or cross-cut boxes at another level. Figure 21 illustrates the Midwestern system employing a smaller number of components and showing the hierarchic relationships between the various levels of units. The similarity of this figure to the dendrogram in Chapter 4 is apparent, this latter construction being the general structure of numerical taxonomies.

The use of the "trait list," especially in subsequent studies employing the Midwestern Taxonomic System, to characterize all of the contents of components and which then serves as the basis for comparing components to state the similarity between

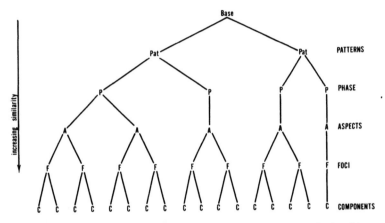

Figure 21. A schematic illustration of the hierarchic relationships within the Midwestern Taxonomic System.

them, presages the polythetic character of modern numerical taxonomy. Importantly, in the 1939 statement of the method McKern emphasizes the phenetic character of the units so formulated. They do not imply "relatedness" or distribution in time and space, but simple formal similarity. This is, of course, a function of choosing number of traits (similarity) over kind of traits as the means of constructing the units. Because there is no control over the kind of criteria used, the resulting units do not have any specifiable meaning. This is important to note, for Figure 21 could be viewed as a taxonomic classification rather than a numerical taxonomy if it were not clear that similarity, not identity, is the basis for construction. Further, as is clear not only from the early formulations of this system but also in its subsequent use, the units consist not of sets of criteria (such is impossible since they can vary from case to case), but of groups of empirical entities, the components.

It has been advantageous to give the Midwestern Taxonomic System detailed and specific consideration because this method is the basis, at least technologically, of all the modern units constructed at the scale of occupation. In subsequent use the higher levels, from aspect upward to base, have gradually been abandoned—generally because chronology and developmental constructions were required, and, one may speculate, because there is no directly analogous unit in sociocultural anthropology above the focus which is equated in a general way with society or "culture." The only major change has been the replacement of the term "focus" with the term "phase" (see Willey and Phillips in bibliography). The notion of "settlement," introduced in recent years by K. C. Chang, the only unit at this scale which departs from the previous formulations, contains strong elements of the Midwestern scheme. Settlement employed as an empirical unit is almost analogous to component and when employed as a class or when the community concerned is not localized in space analogous to focus or phase.

Regardless of the particular terms used in the statement of the system and the number of units retained, there is one important and rather obvious inconsistency—the definitions of phase (focus) and component and the relationship between the two basic units. Components, it is insisted, are empirical units. Yet they are not. Component, regardless of the names used, is a manifestation of a phase or focus at a given locality. This is, of course, a possible way to state the relationship between a class and its *denotata* at a given point in time and space; however, this statement is not a definition by any standards, for if a component is a manifestation of a phase, then one must have phases before one can have components to be able to identify and bound them. On the other hand, phases are said to be groups of components. One must have the components before one can have the phases. This is an interesting circularity, and one which is entirely predictable. As has been pointed out herein in numerous contexts, the "definitions" of groups are always extensional and thus may always be reduced to a statement that "the group is because the group is." The simple fact of the matter is that if one follows the published method, one cannot construct either phases or components; yet, of course, components and phases are constructed. The question thus becomes how.

At the root of the difficulties presented by component/ phase lies the problem of identifying a phenomenological unit larger than discrete object. The solution offered in the literature, calling phases groups of components, is clearly rhetorical and nothing more. The phenomenological units cannot be components, for one cannot identify components without first having phases, and one is still left wondering what phases are units of. Something of a solution is presented in the Midwestern Taxonomic System itself. It was noted that the explication of the system overtly involves numerical taxonomy, but also implicitly, some kind of classification as well. The "traits" used in constructing the units of the system are categorized as linked when shared and as diagnostic when not shared. The diagnostic/linked categorization is exhaustive. A trait is either linked or diagnostic in a given context. There is, however, a third category of traits, a category which is clearly drawn from some system other than the explicated numerical taxonomy, namely determinants. Determinants constitute a set of traits which recur as a complex from component to component and which is distinctive of a focus. Clearly the determinants of a focus constitute a *post hoc* class *significatum*, something quite apart from the system as set forth as a kind of grouping and inconsistent with the system as a whole. Further, since there is no ranking or weighting of the determinant traits, it is reasonable to assume that the determinants of a focus constitute the *significatum* of a paradigmatic class. This is a primary reason that phase was defined as a paradigmatic class of occupations in the previous chapter.

What apparently is generally done by prehistorians, even though explicated in terms of a numerical taxonomy, is paradigmatic classification. This enables one to account for: (1) how it is possible to create phases when the published rationale is insufficient to create them; (2) why only the focus has been seriously retained from the Midwestern Taxonomic System; (3) how it is possible to identify new collections with previously established units; and (4) why determinants, inconsistent with the main theme of the Midwestern Taxonomic System, are nonetheless included in it. This eliminates the circularity of the current treatment of component and phase. Component is used to designate the *denotata* of a class, the phase or focus, at a given locality. Because the *denotata* are real and because the actual units being classified lack discreteness, it has been easy to confuse the results of identification with the phenomena for

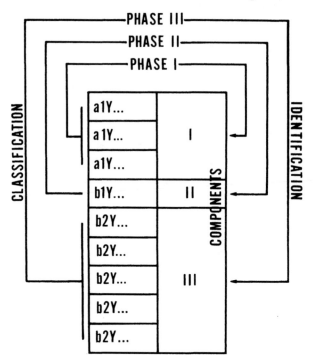

Figure 22. Pragmatic relationships between occupations, phases, and components.

which the classification has been constructed. The relationships between the notions of occupation, phase (focus), and component are illustrated schematically in Figure 22. In this diagram the two columns of boxes represent an ideal stratified site. Those labeled a1y—b2y are occupations, while those labeled I–III represent components. The phases are paradigmatic classes of occupations, the components the *denotata* of each phase. A locality may consist of several occupations all of which belong to the same phase and thus the locality is a single-component site. Alternatively, there may be several occupations which belong to different phases, and thus several components will be recognized at the locality such as in Figure 22.

Viewing the Midwestern Taxonomic System as a numerical taxonomy employed as a rationale for paradigmatic classification eliminates most of the inconsistency in the literature about units at the scale of occupation. It does not, however, eliminate the difficulties inherent in the scale itself. While phases can be treated as paradigmatic classes, components as their *denotata* at a single location, there still is no general definition of what the phenomenological units are. Phases may be paradigmatic classes, but it is not possible to say, at least theoretically, what they are paradigmatic classes of. That such a glaring deficiency should be encountered is not surprising when one considers the dual role the term component has played—on the one hand as the phenomenological unit, and, on the other, the identified *denotata*. Pinpointing the circularity of the component/phase relationship is the crucial first step in correcting this conceptual deficiency.

Other kinds of arrangement, particularly keys, are used from time to time in prehistory. Their use has been rather straightforward, and there is little difficulty in recognizing keys. The only difficulty that inheres in their use is that ordinarily the classification for which the key has been made is not presented separately from the key so that the user is restricted to the classes of the key in his identifications. An excellent example of the key as used in prehistory is included in the appended reading list (see Schwartz, 1961).

Summary

Grouping devices both of the kind herein called statistical clustering and numerical taxonomy occur in the prehistoric literature, and, in fact, constitute some of the better theoretical exposition in the discipline. Both clustering and numerical taxonomy can be done with archaeological materials, but, in spite of lip-service to the contrary, neither has been widely employed in problem-solving for rather simple reasons. Their unit-products are groups, and groups cannot serve as the basis for either comparison or measurement. They are things to be compared and measured. Further, lacking the feature of recurrence necessary for prediction and explanation, their future utility seems unlikely. In the literature, the major uses to which grouping devices have been put are to provide a rationale for paradigmatic classification (inappropriately) and to provide names for the units in "description." Given the inexplicit nature of much prehistoric literature, the identification of the device used to create a set of units is often difficult. In the case of grouping devices, their actual use seems restricted to "descriptive" studies. An ability to distinguish grouping from classification in this context is a moot point; any means of categorizing and naming will suffice, since these kinds of studies have no specifiable problem and thus are not testable. In those cases in which grouping is offered as a rationale for an underlying classification, the use to which the units are put will suffice to indicate the superficial nature of the grouping rationale.

In no fashion is the consideration here any rejection of the techniques of statistical clustering or, for that matter, numerical taxonomy, but only a rejection of their use as means of formulating units. They are exceedingly useful devices for the description of the characteristics of class *denotata* and their behavior in dimensions of variability. They cannot, however, provide useful analytic units for any science. Their appropriate role lies in the generation and testing of hypotheses about classes, not in the construction of the classes. The degree to which grouping can produce usable units is a direct function of the implicit classifications used (attributes and attribute combinations) by these devices. Treating the grouping techniques as the means of unit formation only further obscures the definition of the classes that they must employ.

8

SUMMARY

*T*oday prehistory is rapidly becoming a science, a trend established nearly twenty years ago and one which has gained marked momentum in the last ten. In situations of rapid change there is a strong tendency for an old/new dichotomy to develop and such there is, at least to a moderate degree, in prehistory, represented by what has been called the "old archaeology" and the "new archaeology." The old archaeology, concerned primarily with objects and names for the objects, is giving way to explanatory methods and objectives of the new archaeology. The freedom permitted the workers of the old archaeology as an art is increasingly being constrained by the goals of the new, primarily by the scientific insistence that statements and constructs be testable. The greatest constraints and the greatest progress in reformulating prehistory as a science has been made in the realm of methods of explanation, much of which has been borrowed from and modified after similar procedures of the hard sciences.

The terminological confusion and conceptual imprecision of the old archaeology created few difficulties for the field as an art, for it really did not matter what the terms and concepts meant if they were not to be tested. With the insistence that statements and constructs be testable has come the requirement of knowing precisely what the units and statements mean, and the imprecision of the old archaeology has become a major and, in some respects, almost insurmountable liability. From the point of view of the student of prehistory, the old archaeology could be acquired as an art—by intuitive assimilation. Today the discipline must be treated as a body of knowledge which can be learned. Yet the terminological morass presents serious barriers to such acquisition.

In establishing scientific goals and methods to achieve them, the new archaeology has adopted rather uncritically the units devised by the old archaeology. In many cases, the old units were not specifically designed for any specific purpose, let alone the newly conceived aims of the last two decades. Unfortunately, explanations can be no better than the units they employ and the data they attempt to explain, and the new archaeology has not seriously considered either of these. Indeed, the most serious criticisms of the new archaeology turn on its lack of formal sophistication.

In spite of the inadequacy of the old archaeology when measured by the criteria of science and in spite of the lack of any means of internal evaluation aside from a poll of majority opinion as to what is good or which prehistorian is good, explanations of limited scope and capable of independent evaluation have been made. The question is "how?" What makes the "good" prehistorian good? The old archaeology does not provide explicit answers.

It is in this milieu that we focused upon the construction of units in prehistory, units which have been largely devised by the old archaeology in an unsystematic fashion over a relatively long period of time. These units remain the formal foundation of both new and old archaeologies. Our considerations have been primarily of the old archaeology, which constitutes the bulk of the general literature and the literature on systematics, but with sympathy for the goals of the new. The purpose has been not so much to point out errors and inconsistencies (though this is an important aspect), but to isolate the good features, good, again, from a scientific perspective. Since the many problems faced by the student of prehistory, the confusing terminology, the inconsistencies of method, the lack of evaluative methods, and the lack of any kind of unity of discipline, are all interconnected, a single relatively simple solution was sought in setting forth a unitary system of unit construction including the assumptions on which it is founded and which satisfies the requirements of science. This system was employed to organize and clarify the various units and terms employed by prehistory. Inevitably, much of that currently labeled archaeology was found inadequate as science and discarded from consideration. Much of the literature, however, displays a central theme amenable to interpretation as a scientifically useful system of formal theory. This is nowhere in its entirety spelled out in the archaeological literature. Parts are explicit here, others there, some of it nowhere explicit but implicit only in the operations of the discipline. That a large part of what is done is amenable to such interpretation is a commendation for prehistory. The central theme is essentially an answer to the question of how "good prehistory" and "good prehistorians" have been identified, and why "good prehistory" works. Too, application of the general model permits a unified set of terms, ones which mean the same thing every time they are used, to be employed, and provides a way to identify any unit, given sufficient information, regardless of the name given it in a particular study. A unified terminology was not possible in the old archaeology because of its substantive preoccupation. As long as the names and units were bound to specific studies and specific problems, there was no possibility of developing overt theory. The words for its construction were lacking. In an attempt to avoid this substantive tie and the traditional confusion between concepts and their referents, examples in concrete terms have been kept to a minimum. A unified terminology, measured itself against the goal of science for its utility, can prove of immense benefit in learning prehistory, how it works, its limitations both current and potential, and aid in the selection of units for specific kinds of problems. It is fair, I think, to say that the lack of progress of the old archaeology is not so much a function of the people practicing it or intrinsic failures of the conceptual tools, but of the inconsistent terminology effectively barring much communication and admitting much misunderstanding and error. The rather uncritical borrowing of units by the new archaeology for purposes never conceived when the units were created can be avoided. A general model, no matter how simplified it may be, offers the possibility of distinguishing the inadequacies of theory from the misuse, abuse, and misapplication of good theory. Rational evaluation is possible.

The general model for unit construction in prehistory and

the assumptions it is founded upon have been detailed in the several chapters of Part II in piecemeal fashion in an attempt both to show their rationale in the general scheme presented in Part I and to tie them generally to practice in prehistory. The focus upon the original formulations of classificatory concepts in the discipline rather than more modern renamed and elaborated versions points up the basic lack of change in prehistoric systematics. The remainder of this final chapter attempts to present in brief form this general model, the formal theory of prehistory.

Part I presents the basic notions used throughout, the core of which is the distinction between ideational and phenomenological realms and the correlative distinctions between definition and description, classes and groups, and classification and grouping. The import of the distinctions is two-fold: (1) the means of evaluating ideational constructs and phenomena differ—logical proof in the former case and probability or plausibility in the latter; and (2) the characteristics of ideational constructs and phenomenological units differ in respects that affect their utility in scientific endeavor. Ideational constructs are ahistorical and capable of intensional definition, whereas phenomenological units are contingency-bound and capable of extensional definition at best. The ideational and phenomenological realms are articulated in the framework of science, in that science employs ideational constructs to explain phenomena. The distinction between ideational and phenomenological is entirely analytic, in the nature of the logic of justification. If only because men necessarily use language, the two are fused in reality. The utility of such distinctions lies in their ability to clarify what has been done, for both evaluation and communication, but it is not a program of operational procedures.

Explanation is taken to mean prediction and control. Differences in value are capable of explanation but differences in kind (a function of unit construction) are not. The role of formal theory in science is to provide the means of organizing phenomena so that their explanation is possible. The prime requisite for such organizations is that the units permit recurrence and it is recurrence that enables one to link the known (observed fact) with the unknown (prediction).

Means for unit construction were examined in terms of

the ideational/phenomenological distinction which results in the identification of two fundamentally different devices for creating arrangements—classification, which produces ahistorical, intensionally-defined, ideational units, termed classes; and grouping, which produces contingency-bound, extensionally-defined sets of phenomena called groups. Classes are articulated with phenomena by means of identification, isolating at given points in time and space those phenomena which display the necessary and sufficient conditions for membership as stipulated by the intensional definition. Identified phenomena constitute special groups called the *denotata* of the class. The only groups, aggregates of phenomena, which are capable of explanation in a scientific sense are the *denotata* of classes. Identification is absent with grouping devices since the units and the phenomena are coterminous.

An additional dimension, that of ranking or mutual relationship of units within a system of arrangement, was added to the first distinctions, and two kinds of classification and two kinds of grouping so distinguished: unranked classes all at the same level produced by paradigmatic classification, and hierarchically ranked classes at several levels produced by taxonomic classification; unranked groups all at the same level produced by statistical clustering, and potentially hierarchically ranked groups at several levels produced by numerical taxonomy. Ranking is deemed relevant because it affects the parsimony of the various devices, hierarchic arrangements being the least parsimonious, but most elegant.

A comparison of the four possible means of arrangement with the requirements of unit construction for science identifies classificatory devices as appropriate to this general aim and grouping as inappropriate. Classification produces units which have a defined meaning and the possibility of recurrence, whereas grouping produces units which have no specifiable meaning beyond their existence and which are localized at a given point in time and space. Classification provides a means of measuring changes in value and a basis for comparison. Grouping is not reconcilable to measuring change, and cannot provide the framework for comparison. Within classification, the differences between paradigmatic classification and taxonomic classification affect their utility, or rather their roles in scientific endeavor. Paradigmatic classification, by reason of its relative parsimony, can function as the main tool for unit construction when the purpose is heuristic. Taxonomic classification, because of its lack of parsimony, is useful only in didactive situations in which elegant presentation of already established paradigmatic classifications is required. The characteristic lack of parsimony of taxonomies vitiates their use unless the organization can be shown to be appropriate beforehand.

Another aspect of classification emphasized is the role of problem in its use and evaluation. To construct a classification requires a series of decisions or choices. The field or root of the classification (determined by the discipline), the scale at which features are to be distinguished (determined by the field), and the sets of features to be employed as criteria (determined by the problem) all must be selected and defined. The selection of a feature as definitive represents a hypothesis about the nature of the feature and its relevance for a specific problem. If the hypothesis is assumed to be correct, then the manner in which its use will organize phenomena is predictable. The utility of the hypothesis is tested by comparing the expected distribution with the actual distribution of *denotata*. A given classification either will organize a corpus so that the problem is soluble, or it will not. If alternative classifications are available, the criteria of elegance, parsimony, and sufficiency provide a means of selecting the most appropriate one. Without a specified problem, however, there is no means of justifying the selection of definitive features or evaluating the utility of the classification. Unfortunately, far too often classifications are not accompanied by explicit statements of problem, and this is currently the single greatest operational deficiency in prehistoric systematics.

Grouping, since it produces extensionally defined units, is not amenable to testing. Because groups are applicable as units only to the material from which they are derived, they cannot be tested against new data. Only the mechanics of grouping are testable, not the groups themselves. Groups are thus usually proclaimed as natural or non-arbitrary in lieu of some specified meaning, and the problem for which they serve is that of "description." It was noted that any set of words will suffice the purpose of "description," and thus the means by which the words are invented is irrelevant.

To construct a model of prehistory's formal theory utilizing these notions requires that the discipline be defined as a science in such terms as to delineate empirically the subject matter and to specify the general character of its potential explanations. Prehistory is thus the science of artifacts conducted in terms of the concept culture. Artifact, the concept delineating the subject matter of prehistory, is understood to mean anything which owes any of its attributes to human activity, and culture, the concept controlling the nature of its explanations, is defined as shared ideas. Prehistory is the science which isolates human products and seeks to explain them in human terms. The specification of the kind of study (science), the subject matter (artifact), and the kind of explanation (culture) provides the basic elements for the statement of prehistory's formal theory.

The definitions of both artifact and culture are theoretical and thus do not evidence any particular problem or any particular body of data. They are intended to subsume all of the tactical definitions found in the literature, treating these definitions as special cases of the theoretical definitions and derived by restricting the range of application for a particular problem or body of data. Use of the definitions in the archaeological literature is impossible in the present context because they either evidence some specific content or embody nonessential inferences.

Once the discipline has been defined and the notions of artifact and culture theoretically defined, the development of a model of prehistory's formal operations is relatively straightforward. The aim of the inquiry can be specified. The role of formal theory must be that of creating cultural classifications for artifacts. Making use of the general considerations in Part I, the goal is limited to showing how prehistory converts classification in general into cultural classifications of artifacts. This entailed the identification of the kinds of classification employed and the manner in which they are used. A necessary adjunct is the identification of grouping devices frequently encountered assuming classificatory functions in "descriptive" and "cultural reconstruction" approaches.

The identification of the devices used (classification or grouping) and the forms they take (paradigmatic or taxonomic classification and statistical clustering or numerical taxonomy) is not an easy matter, since the means by which they have been created is rarely explicit, the units customarily taken for granted, and, more often than not, unsystematically named. The account provided for the formal operations of prehistory is thus sufficient only to account for what is done. Its parsimony cannot be assessed directly from the literary sources. The examination suggests that only two kinds of classification are employed: paradigmatic classification, which is by far the most common, and a special form of taxonomic classification which makes use of the dimensional features of paradigmatic classification. True taxonomy is not important; the few cases which are conformable to interpretation as true taxonomy are probably poorly explicated examples of the special-case taxonomy. Both statistical clustering and numerical taxonomy are restricted to the "descriptive" approach with the exception of the general use of numerical taxonomy to provide a rationale for the implicit paradigmatic procedures which lead to the construction of phases. Paradigmatic classes are commonly formed for three scales of phenomena: (1) attributes of discrete objects, with the classes termed "modes"; (2) discrete objects, with the classes termed "types"; and (3) occupations (aggregates of discrete objects), with the classes termed "phases." The special-case taxonomies link alternative paradigmatic classifications which differ from each other in level at all three scales. The number of paradigmatic classifications at any scale is infinitely large, not only by varying the level, but also by changing the criteria used—common means of creating classes to serve specific problems (e.g., functional types, historical types at the scale of discrete object; phases, foci, traditions, horizons, stages, and so forth, at the scale of occupation).

Prehistory has traditionally conceived phenomena in such a manner as to be amenable to scientific explanation. Prehistory further imposes the requirement that the units be cultural. The identities represented by classes and the "behavior" of these classes with respect to other classes and in other dimensions (time/space) must be viewed as the products of ideas held in common by the men who made, used, and deposited the artifacts concerned. How the cultural requirement is met can be treated separately.

Artifacts are identified by the criterion of human involvement and their identification serves to isolate those phenomena amenable to the interests of prehistory. The ease with which artifacts may be identified varies with scale and circumstance;

Summary

a certain number of human products may be excluded because of an inability to reasonably assume their artificial origin. This loss of data is a necessary sacrifice to accuracy. It is essential for the purposes of the discipline that no natural objects be included by its systematics, but it is not equally important that all artificial objects are included. The creation of categories of artifacts requires the stipulation of scale and the three scales enumerated above are nearly universal though not exhaustive in prehistory. Implicit in classification at the lowest scale is a fourth and still lower scale, attributes of attributes of discrete objects. Higher scales make use of classes formed at lower scales in analysis.

Using the concept artifact to segregate phenomena for which cultural classes can be constructed does not mean *ipso facto* that classes of artifacts are cultural. While this possibility is not generally recognized, noncultural classes are in practice generally avoided intuitively. To insure structurally that artifact classes are cultural, additional operations are involved.

First, the potential features themselves must be identified as products of human activity, a parallel procedure to the identification of artifacts and making use of a dichotomous index (artificial attribute/natural attribute). As with artifact, the identification of artificial attributes varies with circumstance, and a certain number of artificial attributes will be lost, since conditions may not reasonably permit the assumption of human involvement. Restricting the source of criteria to artificial attributes insures that artifacts will be treated as human products, eliminating the possibility of natural classes of artifacts.

Secondly, prehistory makes an assumption to convert classes defined on the basis of artificial attributes into cultural classes. This single, simple assumption is the pivotal operation in formal theory, the one upon which the discipline is based— creating features, usually modes, from artificial attributes and completing the articulation of the notion of artifact and that of culture. It is assumed that if a set of objects share the same feature, and that if that feature is artificial, then the objects share that feature because the people responsible shared the same idea. A simple equation is made between recurrent feature and shared idea. As was argued in Part II, this is the only plausible account for shared artificial features. While never explicitly stated in the literature, this is the most universal operation in prehistory, the one which provides such coherence as the discipline has. A certain amount of confusion is possible, and does indeed occur, with regard to this equation: a concern with "intent," contact, and genetic relationships (analogs/homologs). These *post hoc* queries are irrelevant so far as formal theory is concerned. The equation stipulates only formal sharing, that the artifacts concerned may be treated (for a specified purpose) as products of the same mental template. The equation does not say how or why the sharing takes place. This latter concern is entirely inferential; it is an explanation of the distribution of class *denotata,* and it plays no role in their definition.

The testability of classifications is a requirement of science, but a requirement neglected by prehistory even though many of its classifications are testable. The terminological confusion, the vagueness of labeling, and the lack of explicit problems combine to deprive prehistoric classification of much of its potential. Perhaps it is too much to ask for testing and explicit statement of problem when all the preceding operations are implicit, haphazard, and *ad hoc* constructions in common practice. Nonetheless, when the operations are explicit the investigator is forced to make explicit statements of problem and thus creates potentially testable classifications. To construct classes, specific dimensions of features, again usually modes, must be selected as criteria, other dimensions being discarded from a definitive role in the context of a given classification. This decision, or set of decisions, must be justified on the basis of the relevance of the set of dimensions chosen to the problem for which the classes are being constructed. The justification should always have the form of an hypothesis about the nature and relevance of the dimension of modes to the problem. The testing of these devices follows the basic pattern for hypothesis evaluation: the decision, once made, either will create classes which have the necessary distribution characteristics or it will not. An affirmative result (e.g., a particular set of classes will function in seriation) shows that the classes are sufficient. Comparison with alternative classifications makes it possible to ascertain if the classification is the best available.

An important consequence, not as widely recognized and practiced as could be hoped for, is that there are—indeed there must be—as many classifications as there are problems and methods for their solution. Different classifications are created by changing levels (e.g., "ware-type-variety") and by changing criteria (e.g., functional types, ethnological types, historical types). Monolithic "right" classifications have no place in prehistory or any other science which lacks a singularity of inquiry. A single rock will have many names, a single occupation belong to many phases. Our common sense desire for a perfect class/object equation has no utility for phenomena beyond common experience and common interest.

Figure 23 presents in summary fashion the basic elements of formal theory as employed in prehistory. The diagram illustrates only the most usual program, that is, one which begins with modes as the analytic step for the classification of discrete objects. The figure represents the path taken to convert classification into cultural classification of artifacts. The end-product classes, (mode, type, and phase) denote the scale of phenomena for which they are constructed, their paradigmatic nature, and the cultural quality of their defining criteria. These are the only characteristics which can be included for the discipline as a whole in a problem-free context. The number of defining criteria and the particular criteria chosen are a function of the requirements of particular kinds of problems, and variations in these aspects produce the large number of modes, types, and phases, sometimes recognized under special labels such as tradition, horizon, ware, or functional type, and sometimes terminologically undifferentiated. However, prehistory's formal theory, in spite of the presentation afforded it in the literature, is rather simple—there being but three kinds of units employed and their formulation as cultural units of scientific utility being founded upon a small set of simple discriminations and a single, important, but simple assumption.

This view is, of course, not the only one possible. Given the terminological difficulties of prehistory's literature, many accounts are undoubtedly possible. The account presented does offer the advantages of consistency coupled with utility for the purpose of explanation. Also, this is an account of what has been done, not what might be done or could be done. Simply examining Figure 23 suggests that redundancy is involved in the repetition of the artificial/natural discrimination for both artifacts and attributes. This, however, appears to be the way it is done. The prehistorian recovers artifacts and then dis-

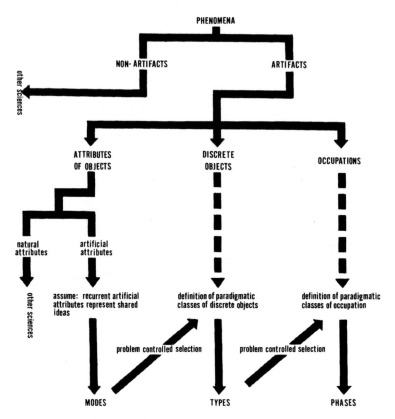

Figure 23.   Elemental formal operations in prehistory.

criminates their attributes rather than recovering artificial attributes and then noting the objects in which they occur.

Larger Issues

Lest the approach taken be misunderstood, it should be clear that the qualitative program adhered to throughout is not a rejection of statistical-oriented approaches. It rather tries to show why grouping and counts are inappropriate to unit construction. The particular view espoused here is perfectly consistent with the goals and most of the methods of the new archaeology, save those attempts to employ groups in the stead of classes. This latter procedure I have tried to show is both counter-productive and deceptive—counter-productive because it does not convey meaning precisely and is not capable of expansion; deceptive because, while it cannot predict, it does provide lucid and elegant accounts of what one already knows.

Statistical techniques have their appropriate role in the manipulation of data, not in its creation. These techniques are necessary to construct distributions and compare and correlate data in their variable aspects, and they will be all the more powerful if they constitute tests of hypotheses embedded in the systematics rather than a shuffling of data. This is possible only when units are properly and explicitly formulated, and it is in this that the "old archaeology" can make an important contribution to the new.

The abuse and inconsistency in the use of classification and its consequent failure in many circumstances certainly points to the need for improvement; however, improvement is not forthcoming from grouping devices. Bad practice has been confused with bad theory and method. The method—classification—is sound, indeed the only device possible, but its practice in prehistory has thwarted its potential. Certainly there are further ramifications. To a formalist, much of the concern with elaborate statistical devices appears as an attempt to correct with a calculus arithmetic errors which have been obscured by multitudinous manipulations. Many of the problems for which factor analysis is proposed, for example, would simply not exist if the formal units had been precisely constructed in the first place with a full knowledge of the wide range of units that can be and in fact have been employed. There may well be a role for factor analysis but it does not appear to be the imperfect correction of errors in systematics; if it has a role to play, this can be reasonably determined only after a reasonable systematic is in use. The qualitative and quantitative are not alternatives; they are necessary complements. The manipulation of the phenomenological world must always be, overtly or covertly, statistical in at least some minimal sense. What correlations mean, not their utility, is questioned. Statistical manipulation is not a means of discovery (this always has been and always will be "guessing"), but a means of testing the efficacy of qualitatively derived units. Regarding the units as givens rather than hypotheses is the principal sin of the old archaeology; currently, disregarding the role of qualitative operations is the sin of the new.

The old archaeology and the new archaeology are not competing approaches, for each has a distinctive and dependent relationship with the other. The old archaeology, basically qualitative, provides the units which the new archaeology can use for explanation. Both are necessary components of a science. The terminological and conceptual ambiguity of the old archaeology resulting in both misapplication of classification and lack of explanation is a function of the state of development of the discipline. There is no model for a science of man, though there are models for parts of it such as have been used here. Errors have been made, perhaps the principal one being the use of sociocultural anthropology as a model. The new archaeology has corrected some of these errors. It has provided the stimulation to correct others. It has incorporated others. The clarification of the old archaeology is a requirement for the success of the new.

DEFINITIONS

1 Preliminary Notions

concept: an intensionally defined term specific to an academic discipline

definition: the necessary and sufficient conditions for membership in a unit

description: a compilation of the variable attributes of an individual case or set of cases

extensional definition: the necessary and sufficient conditions for membership in a unit rendered by enumeration of the members or a statistical summary of same

generalization: a statistical summary of the attributes of a given set of phenomena

hypothesis: a proposed explanation for a specific set of things or events

ideational: anything which does not have objective existence

intensional definition: the necessary and sufficient conditions for membership rendered as a set of distinctive features which an object or event must display to be a member

method: a sub-system of theory which is directed toward the solution of a particular class of problem

phenomenological: anything which has an objective existence

principle (law): a theoretical statement of a relationship held to obtain between two or more classes

science: a systematic study deriving from a logical system which results in the ordering of phenomena to which it is applied in such a manner as to make them ahistorical and capable of explanation (obviously defined to emphasize the role of formal theory)

systematics: the procedures for the creation of sets of units derived from a logical system for a specified purpose

technique: the application of a particular method to a given set of phenomena

theory: a system of units (classes) and relationships (laws) between units that provides the basis for the explanation of phenomena

2   Classification

analysis (analytic step): the discrimination of attributes within a stipulated field and the selection of criteria from such attributes

arbitrary: not inherent in nature as a sole solution

arrangement: any activity which produces ordered sets of units

attribute: the smallest qualitatively distinct unit discriminated for a field of phenomena in a given investigation

class: an intensionally defined unit of meaning

classification: the creation of units of meaning by means of stipulating redundancies

denotatum: any actual instance (thing or event) assigned to a specific class; the means of indicating that an object has been designated as a member of a class

grouping: the creation of units of phenomena

identification: the process of sorting phenomena in terms of class *significata* with the purpose of assigning them to specific classes

significatum: the necessary and sufficient conditions for membership in a class, an intensional definition of a class

3   Kinds of Classification

dimension: a set of mutually exclusive alternative features

index: a unidimensional classification herein treated as a special-case paradigm

paradigmatic classification: dimensional classification in which classes are formed by intersection

Definitions

root: the field of paradigmatic classification expressed as a feature common to all classes in such a classification

taxonomic classification: non-dimensional classification in which classes are defined by means of inclusion

4   Non-Classificatory Arrangement

coefficient of similarity: a numerical expression of the number of features in which two objects or events agree (no scale implied: object may be attribute, discrete object, and so on)

group: an aggregate of actual objects or events, either physically or conceptually associated as a unit (no scale implied)

grouping device: any method for the delineation of units which makes use of quantitative characteristics of a particular set of phenomena to produce units with the characteristics of groups

identification devices: any formal structure designed to assign events or objects to previously defined classes

numerical taxonomy: a grouping device which utilizes similarity of constituent pairs to delimit units

similarity: a quantitative assessment of the number of features shared by two or more objects or events (no scale implied)

statistical clustering: methods of grouping which employ the frequency of association to delimit units

5   Prehistory

artifact: *anything* which exhibits any physical attributes that can be assumed to be the results of human activity

culture: a concept referring to shared ideas used as an explanatory device

prehistory: the science of artifacts and relations between artifacts conducted in terms of culture

6   Classification in Prehistory

data: phenomena categorized for use by a specific science

horizon: a cultural class which displays an extensive distribution in space and a restricted distribution in time (*horizon-style* is applied when the classes are at the scale of attribute)

level: a set of units (classes) which display the same or comparable degree of inclusiveness or rank

mode: (analytic) an intuitive cultural class of attributes of discrete objects; (synthetic) a cultural paradigmatic class of attributes of discrete objects

occupation: a spatial cluster of discrete objects which can reasonably be assumed to be the product of a single group of people at a particular locality and deposited there over a period of continuous residence, comparable to other such units in the same study

phase: (synthetic) a paradigmatic class of occupations defined by types and/or modes

scale: a set of objects (group) which display the same degree of inclusiveness or rank

tradition: a cultural class which displays an extensive distribution in time and a limited distribution in space

type: (analytic) an intuitive cultural class of discrete objects; (synthetic) a paradigmatic class of discrete objects defined by modes

BIBLIOGRAPHY:
GENERAL
SYSTEMATICS

Bayard, Donn T.
   1969 "Science, theory, and reality in the 'New Archaeology.'" *American Antiquity*, 34(4):376–384.
Binford, Lewis R.
   1968 "Archeological perspectives." In Sally R. and Lewis R. Binford, ed., *New Perspectives in Archeology*. Chicago: Aldine. Pp. 5–32.
Brodbeck, May
   1962 "Explanation, prediction, and 'imperfect knowledge.'" In May Brodbeck, ed., *Readings in the Philosophy of the Social Sciences*, 1968. London: Macmillan. Pp. 363–398.
Caldwell, Joseph R.
   1959 "The new American archaeology." *Science,* 129 (3345):303–307.
Clarke, David L.
   1968 "Introduction and polemic." In *Analytical Archaeology*. London: Methuen. Pp. 3–42.
Conklin, Harold C.
   1964 "Ethnogenealogical method." In Ward Goodenough, ed., *Explorations in Cultural Anthropology: Essays*

in Honor of George Peter Murdock. New York: Mc-Graw-Hill. Pp. 25–55.

Gregg, John R.
1964 *The Language of Taxonomy: An Application of Symbolic Logic to the Study of Classificatory Systems.* New York: Columbia University Press.

Hempel, Carl G.
1965 "Fundamentals of taxonomy." In *Aspects of Scientific Explanation and Other Essays in the Philosophy of Science.* New York: The Free Press. Pp. 137–154.
1965 "Typological methods in the natural and social sciences." In *Aspects of Scientific Explanation and Other Essays in the Philosophy of Science.* New York: The Free Press. Pp. 155–171.

Kluckhohn, Clyde
1960 "The use of typology in anthropological theory." In Anthony F. C. Wallace, ed., *Selected Papers of the Fifth International Congress of Anthropological and Ethnological Sciences.* Philadelphia: University of Pennsylvania Press. Pp. 134–140.

Kroeber, A. L.
1940 "Statistical classification." *American Antiquity,* 6(1):29–44.

Lounsbury, Floyd G.
1964 "The structural analysis of kinship semantics." In *Proceedings of the Ninth International Congress of Linguists, Cambridge, Mass., 1962.* The Hague: Mouton. Pp. 1073–1093.

Mayr, Ernst
1961 "Cause and effect in biology." *Science,* 134:1501–1506.

Meehan, Eugene J.
1968 *Explanation in Social Science: A System Paradigm.* Homewood (Illinois): The Dorsey Press.

Meggers, Betty J.
1955 "The coming of age of American archaeology." In *New Interpretations of Aboriginal American Culture, 75th Anniversary Volume of the Anthropological Society of Washington.* Washington, D.C. Pp. 116–129.

Morris, Charles W.
1938 *Foundations of the Theory of Signs.* Foundations of

the Unity of Science, International Encyclopedia of Unified Science, 1(2). Chicago: University of Chicago.

Osborne, Douglas
1968 "Jargon, jabber, and long, long words." *American Antiquity,* 33(3):382–383.

Osgood, Cornelius
1951 "Culture: its empirical and non-empirical character." *Southwestern Journal of Anthropology,* 7:202–214.

Service, Elman R.
1969 "Models for the methodology of mouth-talk." *Southwestern Journal of Anthropology,* 25(1):68–80.

Simpson, George Gaylord
1961 *Principles of Animal Taxonomy.* New York: Columbia University Press.

Sokal, Robert R.
1966 Numerical taxonomy. *Scientific American,* 215(6): 107–117.

Sokal, Robert R., and P. H. A. Sneath
1963 *Principles of Numerical Taxonomy.* San Francisco: W. H. Freeman.

Spaulding, Albert C.
1953 "Statistical techniques for the discovery of artifact types." *American Antiquity,* 18(4):305–313.
1968 Explanation in archeology. In Sally R. and Lewis R. Binford, ed., *New Perspectives in Archeology.* Chicago: Aldine. Pp. 33–39.

Sturtevant, William C.
1964 Studies in Ethnoscience. In A. Kimball Romney and Roy Goodwin D'Andrade, ed., *Transcultural Studies in Cognition. American Anthropologist,* 66(3) part 2:99–131.

BIBLIOGRAPHY: SYSTEMATICS IN PREHISTORY

Benfer, Robert A.
    1967 "A design for the study of archaeological character-
        istics." *American Anthropologist*, 69(6):719–730.
Brew, J. O.
    1946 "The use and abuse of taxonomy." In *Archaeology of
        Alkali Ridge*. Papers of the Peabody Museum of
        Archaeology and Ethnology, Harvard University, No.
        24. Pp. 44–66.
Chang, Kwang-Chih
    1967 *Rethinking Archaeology*. New York: Random House.
    1968 "Toward a science of prehistoric society." In K. C.
        Chang, ed., *Settlement Archaeology*. Palo Alto: Na-
        tional Press. Pp. 1–9.
Deetz, James
    1967 *Invitation to Archaeology*. Garden City: Natural His-
        tory Press.
Doran, James
    1969 "Systems theory, computer simulations and archaeol-
        ogy." *World Archaeology*, 1(3):289–298.
Driver, Harold E.

1965 "Survey of numerical classification in anthropology." In Dell Hymes, ed., *The Use of Computers in Anthropology* (*Studies in General Anthropology*), 2:301–344.

Dunnell, Robert C.
1970 "Seriation method and its evaluation." *American Antiquity*, 35(3):305–319.
1971 Comment on Sabloff and Smith's "The importance of both analytic and taxonomic classification in the type-variety system." *American Antiquity* 36(1): 115–118.

Ford, James A.
1954a Comment on A. C. Spaulding, "Statistical techniques for the discovery of artifact types." *American Antiquity*, 19(4):390–391.
1954b "The type concept revisited." *American Anthropologist*, 56(1):42–54.

Gifford, James C.
1960 "Type-variety method." *American Antiquity*, 25(3): 341–347.

Krieger, Alex D.
1944 "The typological method." *American Antiquity*, 9(3):271–288.

Kroeber, A. L.
1940 "Statistical classification." *American Antiquity*, 6(1):29–44.

Kroeber, A. L., and C. Kluckhohn
1952 *Culture: A Critical Review of Concepts and Definitions*. Papers of the Peabody Museum of American Archaeology and Ethnology, Harvard University, Vol. 47.

McKern, William C.
1939 "The Mid-western taxonomic method as an aid to archaeological study." *American Antiquity*, 4(4): 301–313.

Rands, Robert L.
1961 "Elaboration and invention in ceramic traditions." *American Antiquity*, 26(3):331–340.

Rouse, Irving
1939 *Prehistory in Haiti: A Study in Method.* Yale University Publications in Anthropology, No. 21.

1955 "On the correlation of phases of culture." *American Anthropologist*, 57(4):713–722.

1960 "The classification of artifacts in archaeology." *American Antiquity*, 25(3):313–323.

1968 "Prehistory, typology, and the study of society." In K. C. Chang, ed., *Settlement Archaeology*. Palo Alto: National Press. Pp. 10–30.

Sabloff, Jeremy A., and Robert E. Smith

1969 "The importance of both analytic and taxonomic classification in the type-variety system." *American Antiquity*, 34(3):278–285.

Sackett, James R.

1966 "Quantitative analysis of Upper Paleolithic stone tools." In J. Desmond Clark and F. Clark Howell, ed., *Recent Studies in Paleoanthropology. American Anthropologist*, 68(2, Part 2):356–394.

Schwartz, Douglas W.

1962 "A key to prehistoric Kentucky pottery." *Transactions of the Kentucky Academy of Science*, 22(3–4):82–85.

Spaulding, Albert C.

1953 Review of "Measurements of some prehistoric design developments in the southeastern states," by James A. Ford. *American Anthropologist*, 55(4):588–591.

1953 "Statistical techniques for the discovery of artifact types." *American Antiquity*, 18(4):305–313.

1954 "Reply to Ford" *American Antiquity*, 19(4):391–393.

1960 "Statistical description and comparison of artifact assemblages." In Robert F. Heizer and Sherburne F. Cook, *The Application of Quantitative Methods in Archaeology*. Viking Fund Publications in Anthropology, No. 28. Pp. 60–83.

Steward, Julian H.

1954 "Types of types." *American Anthropologist*, 56(1):54–57.

Taylor, Walter W.

1948 *A Study of Archeology*. American Anthropological Association, memoir 69.

1967 "The sharing criterion and the concept of culture." In Carroll L. Riley and Walter W. Taylor, ed., *American Historical Anthropology: Essays in Honor of*

  Leslie Spier. Carbondale: Southern Illinois University Press. Pp. 221–230.

Wheat, J. B., James C. Gifford, and W. W. Wasley
  1958 "Ceramic variety, type cluster, and ceramic system in Southwestern pottery analysis." *American Antiquity*, 12(4):226–237.

Willey, Gordon R., and Philip Phillips
  1958 *Method and Theory in American Archaeology*. Chicago: University of Chicago Press.

INDEX